Electromagnetic Geothermometry

Electromagnetic Geothermometry

Viacheslav V. Spichak
Olga K. Zakharova

AMSTERDAM • BOSTON • HEIDELBERG • LONDON • NEW YORK
OXFORD • PARIS • SAN DIEGO • SAN FRANCISCO • SINGAPORE
SYDNEY • TOKYO

ELSEVIER

Elsevier
Radarweg 29, PO Box 211, 1000 AE Amsterdam, Netherlands
The Boulevard, Langford Lane, Kidlington, Oxford OX5 1GB, UK
225 Wyman Street, Waltham, MA 02451, USA

Notices
Knowledge and best practice in this field are constantly changing. As new research and experience broaden our understanding, changes in research methods, professional practices, or medical treatment may become necessary.

Practitioners and researchers must always rely on their own experience and knowledge in evaluating and using any information, methods, compounds, or experiments described herein. In using such information or methods they should be mindful of their own safety and the safety of others, including parties for whom they have a professional responsibility.

To the fullest extent of the law, neither the Publisher nor the authors, contributors, or editors, assume any liability for any injury and/or damage to persons or property as a matter of products liability, negligence or otherwise, or from any use or operation of any methods, products, instructions, or ideas contained in the material herein.

British Library Cataloguing-in-Publication Data
A catalogue record for this book is available from the British Library

Library of Congress Cataloging-in-Publication Data
A catalog record for this book is available from the Library of Congress

ISBN: 978-0-12-802210-8

For information on all Elsevier publications
visit our website at http://store.elsevier.com/

Typeset by Thomson Digital

Printed and bound in USA

Working together
to grow libraries in
developing countries

www.elsevier.com • www.bookaid.org

Contents

Part II

Case Studies

Preface

Temperature is one of the key characteristics of the earth's interior, the knowledge of which determines our ability to study both the issues of fundamental science and the applied geothermal problems. Due to this, it appears extremely important to maximally accurately estimate the temperature in the depth interval from a few kilometers (corresponding to the typical borehole depth) to a few dozens of kilometers (corresponding to the depth of the Earth's crust). At the same time, the existing methods of temperature estimation are incapable of providing the required accuracy in this intermediate depth interval and in the cross-borehole space. This difficulty could be overcome by using the so-called proxy parameters depending on temperature (e.g., electrical resistivity of rocks). However, at present, this approach is based on the use of empirical formulas, whose validity is unjustifiably postulated to be invariant with respect to the spatial coordinates.

The progress in this field requires designing of the mathematically sophisticated tools intended for the solution of practical problems. The authors of this book have developed the neural network approach to estimating the temperature in the earth's interior from the data of the ground-based electromagnetic (EM) sounding (the so called Electromagnetic Geothermometer). This method makes it possible to obtain these estimates *in situ* with the accuracy that does not directly depend on the prior information or the assumptions on the physical and chemical properties of the rocks, their metamorphism, lithology, and so on. The purpose of this volume is to provide the methodological basement of EM geothermometry and exemplify its applications for estimating the temperature in the earth interior at different depth scales.

The present book is an expanded English language translation of our monograph "Elektromagnitnii geotermometr," which was published in Russia in 2013. It consists of two parts. In the first one, the survey chapters are followed by the analysis of methodological issues associated with applying of EM geothermometry for interpolation and extrapolation of temperature in the interborehole space and in the geological medium below the boreholes, respectively. In the second part, we consider the application of EM geothermometry for solving the important practical tasks of geothermal exploration in the different geological conditions of the northern Tien Shan (Kyrgyzstan), Travale (Italy), Soultz-sous-Forêts (France), and Hengill (Iceland).

By the examples of these case studies, it is shown that, based on the 2-D and 3-D temperature models, it is possible to draw the conclusions on the location of the heat sources and deep-seated reservoirs, type of fluids, dominant mechanisms of the heat transfer, as well as to constrain the areas for drilling the new boreholes. The material in the second part of the book is presented in the chronological order corresponding to the succession of the investigations that contributed to the experience in applying EM geothermometry. Thus, the discussion of the methodological issues is not limited to the first part of the book.

In our work on this book, we used the material obtained under the international projects INTAS and ENGINE (Sixth European Commission Framework Programme). We acknowledge Drs. S. Bellani, A. Fiordelisi, and A. Rybin, as well as Reykjavik Energy Inc., who provided the temperature data for the Travale-Larderello, northern Tien Shan, and Hengill areas, respectively. We are grateful to Dr. P. Pushkarev who has collected the MT data in the Hengill geothermal zone and to Drs. H. Eysteinsson and K. Árnason, who kindly provided the results of 1-D inversion of TEM data acquired in the Hengill area. We are deeply grateful to Drs. A. Manzella, A. Genter, P. Calcagno, J. Geiermann, and E. Schill, whose close and fruitful cooperation was vital for our work on the thermal projects in Italy and France.

One of the authors (V.V.S.) acknowledges the support from BRGM, GEIE EMC and French Ministry of Foreign Affairs provided for his research at the Soultz-sous-Forêts geothermal site. These studies were also supported by BMU (Germany), ADEME (France), and the consortium of the industrial members (EDF, EnBW, ES, Pfalzwerke, Evonik).

Viacheslav V. Spichak,
Olga K. Zakharova

Moscow, September 10, 2014

Part I

Methodology

Chapter 1

Electromagnetic Sounding of Geothermal Areas

Chapter Outline

1.1 INTRODUCTION

A key issue in the exploration of geothermal systems is the geophysical detection and monitoring, at several kilometers of depth, of reservoirs. Over the past decade there has been a huge increase in time-lapse reservoir monitoring and the development of seismic methods such as repeated 3-D surface seismic, surface-to-borehole vertical seismic profiling, and borehole-to-borehole cross-well seismic. At the same time, electromagnetic (EM) methods have been extensively used to detect deep fluid circulation, since resistivity is very sensitive to the presence of brines. Thanks to improved methodologies and software, EM is now very affordable and logistically practical, and has become very popular. Seismic imaging, while being a powerful geological mapping tool, has not always led to a significant improvement in understanding the nature and composition of the deep structure of geothermal systems. In order to progress and reduce the cost of geothermal exploration and monitoring, resistivity needs to be included in the analysis, especially if it is combined and integrated with other geophysical data. An up-to-date picture of the achievements of EM methods for geothermal exploration will help us to understand and apply modern techniques.

Electromagnetic Geothermometry. http://dx.doi.org/10.1016/B978-0-12-802210-8.00001-0

Geothermal resources are ideal targets for EM methods since they produce strong variations in underground electrical resistivity. Geothermal waters have high concentrations of dissolved salts that result in conducting electrolytes within a rock matrix. The resistivities of both the electrolytes and the rock matrix (to a lesser extent) are temperature dependent in such a way that there is a large reduction in the bulk resistivity to increasing temperatures. The resulting resistivity is also related to the presence of clay minerals and can be reduced considerably when clay minerals and clay-sized particles are broadly distributed. On the other hand, resistivity should be always considered with care. Experience has shown that the correlation between low resistivity and fluid concentration is not always correct since alteration minerals produce comparable and often a greater reduction in resistivity. Moreover, although water-dominated geothermal systems have an associated low-resistivity signature, the opposite is not true, and the analysis requires the inclusion of other geophysical data in order to limit the uncertainties.

Many papers have been devoted to the study of geothermal areas by EM methods for the last 30 years (see review papers by Berktold (1983), Meju (2002), Munõz (2014), and references therein). Recently a number of important achievements have been reported, especially in EM data interpretation, and they are reviewed in this chapter following (Spichak and Manzella, 2009). First we will summarize the conceptual models of geothermal areas, the main factors influencing rock resistivity, and how they are evaluated using EM data. We will then present the results of applying EM techniques to geothermal targets, with particular focus on magnetotelluric (MT) techniques but also looking at other EM methods. We will consider different aspects of EM data interpretation, with emphasis on new approaches of 3-D data inversion. A special section will be devoted to the effects of fracturing, faulting, and regional tectonics on the detectability of geothermal zones using EM methods. We will also discuss MT monitoring of the reservoir macroparameters and methods for dealing with cultural and geological noises. We will address modern techniques of joint analysis and inversion of EM and other geophysical data, as well as the important practical problem of defining drilling targets depending on the type of geothermal zone. Finally, we will outline the latest contribution of EM sounding to geothermal exploration and the direction of future developments.

1.2   CONCEPTUAL MODELS OF GEOTHERMAL AREAS

Geothermal resources are often confused with hydrothermal systems. By the latter we mean large amounts of hot, natural fluids contained in fractures and pores within rocks at temperatures above ambient level. Typically, when fluids are tapped at the surface either by natural manifestation or through drilling, hot water or steam is produced and its energy is converted into marketable products (electricity, heat). A hydrothermal system is made up of three main elements: a heat source (very often represented by a magma chamber or intrusive bodies), a reservoir (i.e., a constituent host rock and the natural fluids contained in its fractures and pores), and a cap rock (i.e., a low permeability layer, which restrains the main fluid flow at a depth where the temperature is high and is prevented from cooling by mixing with surface water). The sustainability of the system is guaranteed only when sufficient recharge through meteoric water is available, usually at a certain distance from the main hot fluid circulation.

Geothermal resources refer to the thermal energy stored in the earth's crust. For many tens of years, the geothermal community has tried to broaden the categories of geothermal systems beyond economically viable hydrothermal systems. The term "enhanced geothermal system" is used nowadays to classify low permeability/porosity rock volumes at high temperatures that are stimulated (i.e., fractured) to extract economically justified amounts of heat. Another important frontier in geothermal research is linked to rocks that contain fluid in supercritical conditions, for which the conversion from thermal energy to mechanical energy would be particularly efficient. These different classes of geothermal resources have one parameter in common: temperature. Hence, the primary aim of geothermal exploration is to map the temperature and heat. If there is a reasonable temperature at depth, geothermal explorers should be able to define the mineralogical composition of rocks, rheological conditions, but they are particularly interested in fluid pathways. All the aspects described so far have a direct effect on the resistivity distribution at depth.

Geothermal explorations including EM methods have mainly been carried out in hydrothermal systems. Modern geothermal exploration, however, should be able to distinguish between different kinds of situations. The main difference between hydrothermal systems and other classes of geothermal resources is the rate of rock alteration, since hydrothermal systems are characterized by a prolonged water–rock interaction effect. Apart from this aspect, most of the following review, which refers primarily to hydrothermal systems, may be applied to any geothermal system.

In geothermal areas where the permeability is high and alteration pervasive, the conceptual model of the reservoir shown in Figure 1.1 is appropriate. Reservoirs of this type have been found, for example, in Iceland, New Zealand, El Salvador, Djibouti, Indonesia, and Japan (Árnason et al., 1986, 2000; Árnasson and Flóvenz, 1995; Uchida, 1995; Oskooi et al., 2005). In this model, the lowest resistivity corresponds to a clay cap overlying the geothermal reservoir, while the resistivity of the reservoir itself may be much higher.

When topography is steep and a significant hydrological gradient is present in the subsurface, the overall structure of the geothermal system is more complex (Figure 1.2). The conductive clay layer, for example, smectite, may be quite deep over the system upflow and much closer to the surface in cooler outflow areas. In these cases, the resistivity anomaly at the surface is not centered over the geothermal reservoir (Anderson et al., 2000).

High-temperature geothermal systems, which are required for electrical power production, usually occur where magma intrudes into high crustal levels (<10 km) and hydrothermal convection can take place above the intrusive body (e.g., Ander et al., 1984; Mogi and Nakama, 1993; Bai et al., 2001; Ushijima

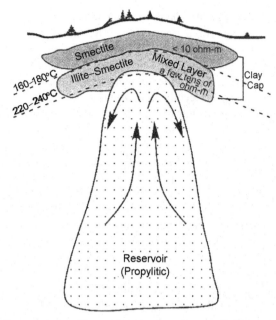

FIGURE 1.1 Conceptual resistivity model of a convective geothermal system (after Oskooi et al., 2005).

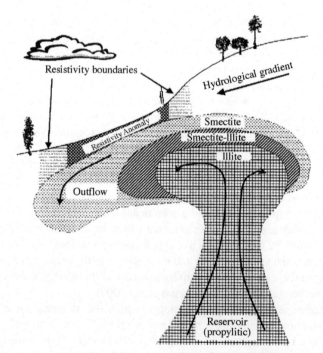

FIGURE 1.2 A generalized geothermal system in a steep terrain (after Anderson et al., 2000).

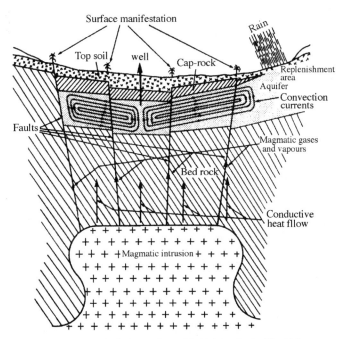

FIGURE 1.3 Conceptual model of a hyper-thermal field (after Berktold, 1983).

et al., 2005; Veeraswamy and Harinarayana, 2006; Zlotnicki et al., 2006). Figure 1.3 indicates a conceptual model showing the main elements of this type of geothermal system (Berktold, 1983).

1.3 FACTORS AFFECTING ELECTRICAL RESISTIVITY OF ROCKS

EM, and in particular MT sounding, is a direct method for *in situ* studies of the structure and fluid circulation in the earth's crust. Due to relations between the electrical resistivity, on the one hand, and temperature, porosity, permeability, and alteration mineralogy, on the other hand, it is often used for the indirect estimation of these parameters.

1.3.1 Temperature

Resistivity in hydrothermal areas is affected by vertically ascending, hot mineralized waters or gases that originate from the contact between groundwater and high temperature intrusive magmas. The intrusions themselves have a very low intrinsic resistivity at temperatures above approximately 800°C (Bartel and Jacobson, 1987). The resistivity of a solid phase correlates with temperature and is documented in the laboratory through the semiconductor equation:

$$\rho = \rho_0 \, e^{E/kT}, \tag{1.1}$$

where ρ_0 is the resistivity at a theoretically infinite temperature, E is an activation energy, k is a Boltzman constant, and T is an absolute temperature (K). However, when laboratory parameters and reasonable temperatures are used in the relation (1.1), it often does not yield resistivities as small as those obtained from field measurements. In order to fix this problem Flóvenz et al. (1985) use the following empirical relation between the resistivity of altered basalts and temperature:

$$\rho_T = \rho_0 / ((1+\alpha(T-T_0))((1+\beta(T-T_0)), \tag{1.2}$$

where ρ_T and ρ_0 are the resistivities at temperature T and reference temperature T_0. Empirical constants α and β are determined at T_0. This empirical relationship for resistivity exhibits about twice as much temperature dependency compared with electrolytic conduction alone. Flóvenz and Karlsdottir (2000) studied the dependence of the resistivity of the low-resistivity layer revealed by a transient electromagnetic (TEM) survey on the temperature estimated from borehole temperature logs. The authors observed a good correlation between resistivity and temperature in agreement with the empirical relation (1.2) and explain this by interface conduction along very thin but highly conductive films of clay minerals covering the interconnected microcracks in the rock.

1.3.2   Rock Porosity and Permeability

A quantitative empirical relation between the electrical conductivity (the reciprocal of resistivity) of rocks and their porosity was first established by Archie (1942). A later formulation by Waxman and Smith (1968) allows for an additional conductivity contribution called surface conductivity σ_s:

$$\sigma_r = (1/F)\sigma_w + \sigma_s, \tag{1.3}$$

where σ_r is the bulk electrical conductivity of a rock, σ_w is the electrical conductivity of a saturating fluid, and F is the formation factor of the sample. In practice, if the ionic concentration of the solution is higher than 0.01M, the contribution of surface conductivity in (1.3) can be neglected (Shmonov et al., 2000).

The formation factor F was expressed by Archie (1942) in terms of the fractional porosity ϕ as $F = \phi^{-m}$, where m is the cementation factor varying from 1.4 to 2.2 (Ward, 1990). It is often taken to be equal to 2. With this substitution the electrical conductivity of a fluid-saturated rock is approximately

$$\sigma_r = \sigma_w \phi^2 \tag{1.4}$$

If the interconnected pore spaces are not damaged in the course of rock deformation, the hydrologic permeability k can be expressed by the relation (Zhang et al., 1994)

$$k = \alpha\phi^3 \tag{1.5}$$

Coefficient α in equation (1.5) has a value of $1 \times 10^{-12}\,m^2$. Combining (1.4) and (1.5), the electrical conductivity of a rock can be estimated from the electrical conductivity of a solution and the hydrologic permeability from the relation

$$\sigma_r = \sigma_w \left(k/\alpha\right)^{2/3} \tag{1.6}$$

Electrical conductivity of a fluid increases as a function of temperature and pressure, whereas permeability can both decrease and increase with respect to these variables. But at low and moderate temperatures the expected variations in k or σ do not result in sharp conductivity increases with depth. In fact, at depths of more than 15 km the rock electrical conductivity often decreases. This occurs because the permeability of rocks decreases due to compaction and temperature-dependent plasticity (Shmonov et al., 2000).

Electrical conductivity values of the order of 10^{-1} S/m can be reached in regions with high heat flows at depths of about 20 km where temperatures higher than 500 °C occur (Shmonov et al., 2000) (Figure 1.4). In this case electrical conductivity changes from $\sim$0.001 to 0.05 S/m are expected at depths from 20 to 25 km (e.g., rock sample 31863p in Figure 1.4). Such a rapid increase in electrical conductivity ($\sim$0.0116 S/m km) produces "conductivity jumps" that provide excellent targets for EM soundings.

Evaluation of rock porosity and permeability from EM sounding data would significantly facilitate geothermal exploration and reduce drilling costs. Svetov

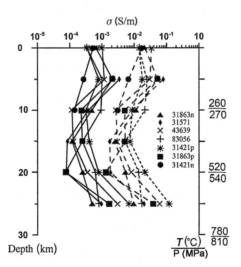

FIGURE 1.4 Calculated values of electrical conductivity of rocks with a fluid represented by a 0.1M solution of NaCl (symbols are linked by solid lines) and 3.3M solution of KCl (symbols are linked by dash lines) for regions with high heat flows (temperature gradient 26 °C/km) (after Shmonov et al., 2000). Electrical conductivity is given in logarithmic scale. Symbols "p" and "n" at rock sample numbers indicate that in calculations the data on permeability in the directions parallel and perpendicular to the bedding, respectively, were used.

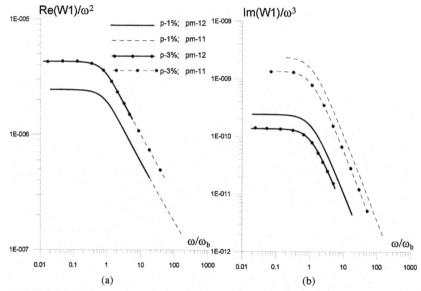

FIGURE 1.5   Frequency dependences of real (a) and imaginary (b) parts of seismoelectric transfer function W1 for different values of porosity and permeability; p stands for porosity and pm for permeability (after Svetov, 2006).

(2006) has shown that seismoelectrical transfer functions calculated from the EM data, measured at the earth's surface or in the borehole, could be sensitive to the porosity and permeability of the earth's crust.

Figure 1.5 shows frequency characteristics of real and imaginary parts of seismoelectrical transfer function (normalized by resistivity of a homogeneous crust) calculated for a longitudinal wave. The real part is additionally normalized by ω^2, and the imaginary by ω^3 (where ω is the EM field frequency). At not very high frequencies ($\omega \ll \omega_b$), the real part is proportional to porosity and is practically permeability independent (Figure 1.5a), whereas the imaginary part depends only weakly on porosity and is proportional to permeability (Figure 1.5b).

1.3.3   Alteration Mineralogy

Resistivity in geothermal areas is also governed to a great extent by the presence of hydrothermal alteration products including clay minerals. Clays are found in natural environments due to low-grade metamorphic and hydrothermal conditions. Systematic petrological, mineralogical, chemical, and well-logging studies have been performed in recent years to investigate different hydrothermal alteration zones from the surface to great depths in geothermal fields worldwide.

Temperature is the major control of clay mineralogy. Below the cool, unaltered shallow part of the earth, the environment is characterized by low-temperature clay minerals such as smectite and zeolites. Both are electrically conductive and form at temperatures above 70°C. At higher temperatures, chlorite (more abundant in basaltic rocks) and/or illite (a less conductive clay mineral in acidic rocks) may appear, interlayered with low-temperature alteration minerals. The proportion of chlorite or illite increases with temperature, especially above 180°C. Zeolites and smectite disappear at 220–240°C and pure chlorite and/or illite usually appear at temperatures higher than 240°C, along with other high temperature alteration minerals such as epidote in propylitic alteration assemblages (e.g., Patrier et al., 1996; Fulignati et al., 1997; Gonzalez-Partida et al., 1997; Gianelli et al., 1998; Srodon, 1999; Lackschewitz et al., 2000; Gonzalez-Partida et al., 2000; Okada et al., 2000; Sener and Gervek, 2000; Yang et al., 2000; Suharno et al., 2000; Natland and Dick, 2001; Yang et al., 2001).

If a rock contains clay minerals, then an extra conduction pathway is possible via the electrical double layer that forms at the interface of the clay mineral and the water (for a more detailed discussion see Ward (1990)). This allows ions to move through the system with a lower effective viscosity than in the liquid phase. Such an additional contribution to the conductivity is referred to as surface conductivity or the double-layer effect. It can exceed the conductivity of the water itself by many times and is thus of utmost importance in clay-rich zones.

Ussher et al. (2000) discussed the effect of clay minerals on surface electrical conductivity. According to the authors, there are two conduction paths through clay-rich sediments: (1) via pore water and (2) a double layer of cations (also called the Gouy layer), which occurs in water at the interface of clay minerals. An analogue of this is an electrical circuit with two parallel paths as expressed by (1.3).

In general surface electrical conductivity is determined as

$$\sigma_S = B Q_V, \tag{1.7}$$

where B is the equivalent conductance of counter ions (a function of solution conductivity), and Q_V is defined by

$$Q_V = CEC(1-\phi)\rho_m\phi^{-1} \tag{1.8}$$

where ρ_m is matrix grain density and CEC is the cation exchange capacity of the clay present (in meq/g).

In clay-rich rocks and where the pore water has low salinity, the bulk electrical conductivity of the rock is proportional to the CEC of the clays. The role of this conduction mechanism in geothermal exploration was studied by Flóvenz et al. (2005). The authors conclude that for almost all freshwater-saturated high-temperature geothermal fields interface conduction is the dominant conduction mechanism, both in the chlorite and smectite zones. Thus, the observed decrease in resistivity at the top of a mixed clay/chlorite zone in many high-temperature geothermal reservoirs worldwide is most likely due to a much higher cation exchange capacity in smectites than in the alteration minerals below.

The effect of surface conductivity is generally more important when the water conductivity; therefore, the intrinsic ion content or the porosity or permeability is low. For example, when permeability is supported by a few main fractures and faults, such as in the Amiata area in Italy, then alteration minerals are localized, and the decrease in resistivity is mainly caused by the presence of hydrothermal fluids and partial melt (Volpi et al., 2003).

Alteration minerals produce comparable or even greater reductions in resistivity with respect to fluid flow. Resistivity distribution provides a clear indication of the presence of hydrothermal alteration minerals and is often considered an important tool to locate geothermal reservoirs. However, it is difficult to directly determine the alteration mineralogy from EM sounding data, since different mineral assemblages may show the same resistivity.

On the other hand, the temperature dependence of the alteration mineralogy makes it possible to interpret the resistivity layering in terms of temperature, provided the temperature is in equilibrium with the dominant alteration (Árnason et al., 2000). According to these authors, the upper boundary of the low-resistivity cap corresponds to temperatures in the range of 50–100°C, depending on the intensity of the alteration. The transition from the low-resistivity cap to the resistive core corresponds to temperatures in the range of 230–250°C. Thus, if alteration is in equilibrium with temperature, the mapping of the resistivity structure is in fact the mapping of isotherms.

Anderson et al. (2000) proposed a simple imaging of the high-temperature geothermal reservoirs by mapping the base of the conductive layer on the assumption that it corresponds to the 180°C isotherm. However, resistivity reflects the alteration but not the present temperature, if cooling has recently taken place. In this case resistivity should be considered as a maximum geothermometer (Árnason et al., 2000). In order to detect whether the geothermal system is still active, resistivity data should be supplemented by temperature estimations (see in this relation Section 8.6 of the Chapter 8).

1.4 IMAGING OF GEOTHERMAL AREAS

1.4.1 MT Sounding

Of the various EM methods, MT was found to be the most effective in defining a conductive reservoir overlain by a larger and more conductive clay cap. This is because the main anomaly related to the reservoir is caused by the presence of electrical charges at conductivity boundaries rather than EM induction, and methods measuring the electrical field are superior to techniques that only employ magnetic field measurements (Pellerin et al., 1996). The depth of investigation of MT is much greater at long source periods compared to most controlled-source EM methods. Such methods (see the next section of this chapter) are usually unable to detect geothermal reservoirs deeper than

1–2 km. The natural-field MT method has also proved very useful for mapping near-surface low-resistivity zones caused by rock alteration and saline pore fluids in geothermal areas (e.g., Hoover et al., 1976; Hoover et al., 1978; Long and Kaufman, 1980; Ushijima et al., 1986; Haak et al., 1989; Ingham, 1991; Anderson et al., 1995; Uchida, 1995; Lagios et al., 1998; Mulyadi, 2000; Uchida et al., 2000; Bai et al., 2001; Hafizi et al., 2002; Caglar et al., 2005; Correia and Safanda, 2002; Risk et al., 2002; Pérez-Flores and Schultz, 2002; Wannamaker et al., 2002; Wannamaker et al., 2005; Volpi et al., 2003; Ushijima et al., 2005).

MT data inversion using geological and geophysical constraints enables not only to determine the general conductivity distribution in the studied area but also to delineate the geothermal reservoir and detect the cap layer. In particular, Uchida (1995) inverted the MT data in the Sumikawa geothermal field of northeastern Japan and interpreted a two-dimensional (2-D) resistivity model using drilling constraints such as temperature, porosity, clay mineral content, and resistivity logs. The resulting resistivity model has two major features: (1) a cap layer of low resistivity (1–3 Ωm) due to low temperature clay minerals and (2) a more resistive reservoir (~100 Ωm) in spite of its high temperature.

MT sounding enables also to indicate the location of permeable and fluid-saturated areas. In particular, Caglar et al. (2005) presented an MT survey in the Afyon part of the Taride zone in southwest Anatolia. Models of the resistivity structures obtained from 2-D inversion of MT data clearly showed the existence of an electrically conductive (<50 Ωm) zone beneath the Sandikli graben geothermal region. Another conductive zone with resistivity values of 5–10 Ωm was defined beneath the seismically active Dinar graben. The authors believe that the very low-resistivity values probably suggest that both zones are highly permeable and saturated with geothermal fluids.

MT soundings of volcanic geothermal areas enable to detect the location of the geothermal reservoir beneath the volcanic sediments (e.g., Ander et al., 1984; Eysteinsson and Hermance, 1985; Flóvenz et al., 1985; Hermance, 1985; Bartel and Jacobson, 1987; Mogi and Nakama, 1993; Bai et al., 2001; Talebi et al., 2005; Ushijima et al., 2005; Spichak et al., 2007b). In particular, Ushijima et al. (2005) used MT sounding to reveal anomalously low-resistivity zones in the Takigami geothermal area, Kyushu, Japan, indicating a potential geothermal reservoir beneath the volcanic area. The low-resistivity patterns often correlate with the distribution of smectite clay alteration products and are generally consistent with higher temperatures revealed from exploration wells (Talebi et al., 2005) (Figure 1.6).

Thus, MT imaging of geothermal zones provides information on the base of the conductive clay cap and spatial location of the anomalously low-resistivity zones which could be considered as candidates for being the geothermal reservoir. A comparison of the resistivity section with temperature distribution and alteration mineralogy enhances the reliability of MT data interpretation in geothermal terms.

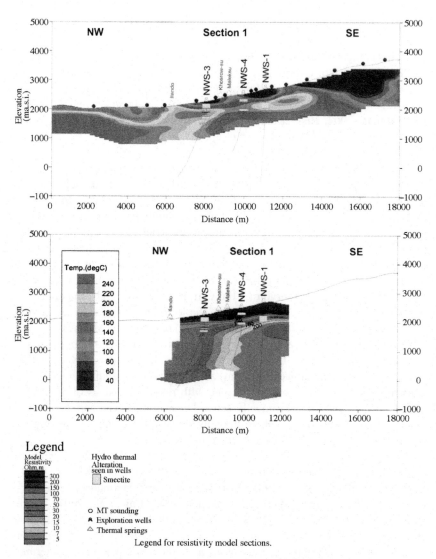

FIGURE 1.6 2-D resistivity section along the profile 1 (top) and corresponding temperature distribution (bottom) revealed from exploration wells (after Talebi et al., 2005).

1.4.2 Other EM Methods

In areas where the geothermal circulation and related alteration take place at shallow depths (<1 km), it is better to use TEM or direct current (DC) methods. Bibby et al. (1995), Árnason et al. (2000), and Pérez-Flores et al. (2005) used DC (Schlumberger) sounding of the geothermal areas located in the Taupo volcanic zone (TVZ) of New Zealand, Iceland, and the Cerro Prieto region of

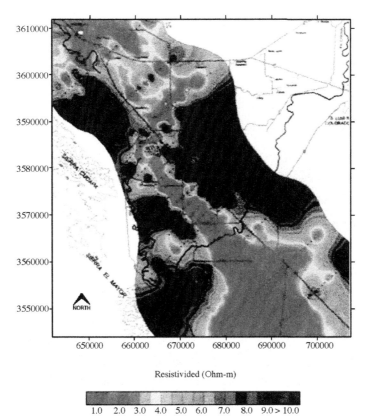

Resistivided (Ohm-m)

1.0 2.0 3.0 4.0 5.0 6.0 7.0 8.0 9.0 > 10.0

FIGURE 1.7 Horizontal slice of the 3-D resistivity model obtained at 1200 m depth. Three linear trends seem to be present, two of them correspond to Cerro Prieto and Imperial faults (after Pérez-Flores et al., 2005).

Mexico, respectively, while Pérez-Flores and Gomez-Treviño (1997) applied DC (dipole–dipole) resistivity imaging of the Ahuachapan-Chipilapa (El Salvador) geothermal field.

In particular, the Cerro Prieto region was prospected with more than 400 long offset Schlumberger soundings. The geothermal area is at the center of a system of echelon faults that produce a slimming and possible rupture of the earth's crust. With the help of resistivity data, the authors obtained a 3-D resistivity image of the geothermal area and of the two principal faults that control the regional tectonics (Figure 1.7). The application of fast imaging techniques developed by Pérez-Flores and Gomez-Treviño (1997) to dipole–dipole resistivity data in the Ahuachapan-Chipilapa area resulted in resistivity images that correlate with surface hydrothermal manifestations, information derived from drilled wells, and the results of MT surveys.

In order to reduce the negative effect of the static shift on the results of MT soundings it is common practice to correct the near-surface resistivity values with results of DC and/or TEM surveys (Ushijima et al., 1986; Bromley, 1993; Gunderson et al., 2000; Romo et al., 2000; Kajiwara et al., 2000) (see also Section 8.3.1 in Chapter 8). Árnason et al. (2000), Flóvenz and Karlsdottir (2000), and Demissie (2005) have reported on TEM soundings in Iceland. In the Eyjafjordur geothermal zone (northern Iceland), TEM resistivity soundings have been used by Flóvenz and Karlsdottir (2000) to create a resistivity image of the basaltic pile in the area to improve the older resistivity picture obtained by Schlumberger soundings. Inversion of the TEM soundings resulted in a layered resistivity model consisting of several layers of high and low resistivity with a southerly dip. This pattern had not been resolved in the previous Schlumberger soundings. A comparison of the TEM result with borehole and surface data shows that the resistivity layering coincides with the lithological one.

Heat-triggering phenomena (such as thermoelectrical and electrokinetic effects) due to hydrothermal circulation are manifested in self-potential anomalies observed at the earth's surface. Results obtained by Zlotnicki et al. (2006) show that self-potential anomalies could be an efficient indicator of the change in the thermal state and evolution of hydrothermal activity.

It is worth mentioning the method of resistivity and permittivity reconstruction from multifrequency EM well-logging data proposed by Shen et al. (2000). It may improve the resolution of EM sounding in low-resistivity contrast zones (especially in areas where the water injection technique is carried out).

1.4.3 3-D Resistivity Models of Geothermal Zones

The geological structure of geothermal areas is often very complicated due to hydrothermal circulation and alteration. Because of this, 1-D and 2-D interpretations of EM data may result in erroneous conclusions regarding the location and boundaries of the geothermal reservoir (especially in deep parts). Moreover, only 3-D models of the geothermal areas enable to guide the best locations for future drilling. 3-D modeling, imaging, and inversion tools currently available enable to reconstruct the most adequate geoelectrical structures of geothermal areas, which, in turn, could be used to estimate the volcano's energy (Mogi and Nakama, 1993), delineate the conductive clay cap (Spichak, 2002) (Figure 1.8) and geothermal reservoirs (Uchida, 2005; Uchida et al., 2005) (Figure 1.9). By comparing the resistivity distributions with the temperature distributions based on fluid-flow calculations at a steady state, the validity of the location and dimensions of the estimated reservoirs could be confirmed (Asaue et al., 2006).

Thus, the results discussed above clearly demonstrate the potential of the MT method in spatial mapping geothermal zones. 2-D interpretation of the MT data in such complicated areas as geothermal zones is often insufficient. Modern 3-D inversion methods provide more appropriate tools for adequate resistivity reconstruction and, thus, for a more reliable estimation of the energetic potential of the reservoir and guiding locations for future drillings.

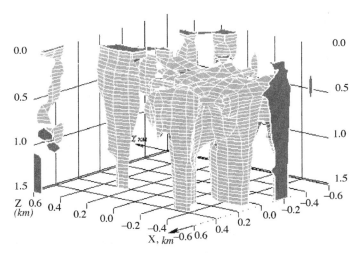

FIGURE 1.8 Highly conductive cap (resistivity <6 Ωm) in the Minamikayabe geothermal zone (after Spichak, 2002).

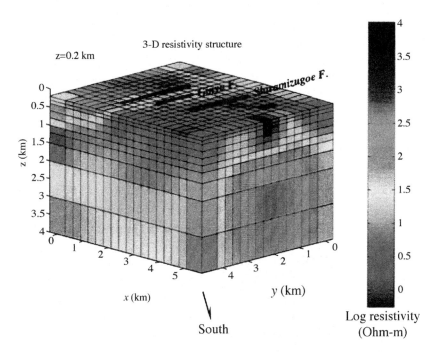

FIGURE 1.9 3-D view of the resistivity model of the Ogiri geothermal zone, Japan, from south. Shallow blocks to a 200-m depth are stripped out and approximate locations of three faults are overlaid (after Uchida, 2005).

1.5   EM FOOTPRINTS OF THERMOTECTONICS, FAULTING, AND FRACTURING

In terms of plate tectonics, geothermal regions occur mainly at or near-divergent and convergent plate boundaries (spreading ridges and subduction zones), at intraplate rifts (continental rifts and thermal anomalies [hot spots]). Plate movement may be accompanied by magma intrusions into the crust and by volcanism. Tectonic activity may cause deep faults and fractures in the sedimentary cover and in the upper crust. Figure 1.10 shows junctions of mobile belts and anomalous thermotectonics (circles with cross) along with the heat flow values in different regions of India (Veeraswamy and Harinarayana, 2006). In these areas heat is transferred from the earth's interior not only by conduction but also by vertical mass transfer, such as upwelling of magma or deep water circulation and heat flow is often increased regionally with local extrema.

The presence of high-conductive anomalies in the upper crust of the Himalayan belt region is attributed to the presence of partial melt generated from the subducted Indian crust. Harinarayana and Veeraswamy (Harinarayana et al., 2006; Veeraswamy and Harinarayana, 2006) studied the Puga and Tatapani geothermal zones, located in the tectonically active regions of India, using 2-D inversion of MT data. The subsurface sections along three E–W oriented

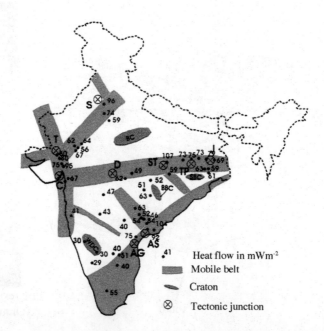

FIGURE 1.10   Junctions of mobile belts and anomalous thermotectonics (circles with cross) along with the heat flow values in different regions of India. C: Cambay; TP: Tatapani, D: Damua/ Parsia, J: Jharia, ST: Sonipat/Tusham, T: Tharad, AG: Agnigundala, AS: Aswaraopeta, and others (after Veeraswamy and Harinarayana, 2006).

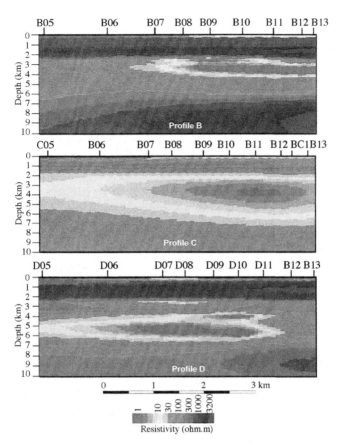

FIGURE 1.11 Resistivity cross-section obtained from the 2-D inversion of magnetotelluric data along three profile lines (B (top), C (middle) and D (bottom)) in the Puga geothermal area, India (after Harinarayana et al., 2006).

profiles (Figure 1.11) indicated the presence of an anomalous high-conductivity zone up to a depth of about 300–400 m near the center of the profile. Such an area is also evident at a depth of about 1.5–2 km. The deeper anomalous feature is clearly delineated with a width of about 4 km indicating its greater size, which probably represents the geothermal reservoir. Maybe, the most useful way to get an idea about local and regional tectonics, which is related to the subduction and distensional processes, is to provide analysis of azimuths of the major axes of the MT impedance ellipses and Parkinson induction arrows (Galanopoulus et al., 1991).

Rock fracturing is necessary for a promising reservoir and for the maintenance of well capacities throughout the life of the geothermal power plant. It leads to an increase in rock permeability, which, in turn, provides favorable conditions for hydrothermal fluid circulation. The latter is often detected

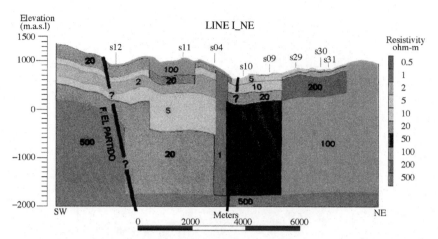

FIGURE 1.12   Subsurface resistivity model of the Las Tres Virgenes geothermal field (Baya California Peninsula, Mexico) along line I_NE (after Romo et al., 2000).

along faults; thus, EM sounding of faulted zones may help to guide the location of geothermal reservoirs (e.g., Ushijima et al., 1986; Galanopoulus et al., 1991; Bromley, 1993; Bibby et al., 1995; Correia and Jones, 1997; Wannamaker, 1997a,b; Lagios et al., 1998; Romo et al., 2000; De Lugao et al., 2002; Wannamaker et al., 2002; Del Rosario et al., 2005; Layugan et al., 2005; Manzella et al., 2006).

Romo et al. (2000) used MT measurements at 90 sites to estimate the subsurface distribution of the electrical conductivity at a depth range from 0 to 3 km. The results suggest the presence of a highly attenuating and conductive zone along the El Azufre Canyon (the boundary between Las Tres Virgenes complex and El Aguajito complex), which corresponds to the production interval of wells LV-2 and LV-3/4. A graben structure (Figure 1.12), outlined between sites s12 and s10, is bounded by 1 Ωm vertical conductors presumably associated with the El Partido and El Azufre faults.

In contrary to the above case the geothermal areas located in andesitic volcano–sedimentary systems (Layugan et al., 2005; Los Banos and Maneja, 2005) are often more resistive than surrounding sediments being caused by high-temperature but less conductive minerals such as illite, epidote, actinolite, and garnet developed within the sediments. Figure 1.13 shows the 2-D resistivity model of the Southern Leyte geothermal field (Philippine) as an example. The intrusion-like resistive body (>40–100 Ωm) beneath the western area of Mt. Cantoyocdoc is interpreted by the authors as the hottest region of the geothermal system. The high-resistivity area is bounded by moderate resistivity values (>20–40 Ωm) that most likely reflect the sedimentary rock situated away from the center of the resource, where geothermal activity is absent.

Wannamaker (1997b) showed the advantages of having long period MT data combined with controlled-source audio-magnetotelluric data when surveying the

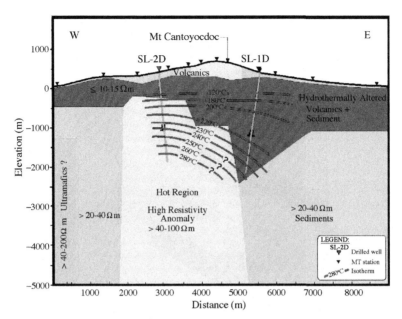

FIGURE 1.13 MT resistivity structure correlated with lithology and temperature at Southern Leyte, Philippines (after Layugan et al., 2005).

characteristics of the Sulphur Springs thermal area in the United States. Parameterized 2-D inversion of this data resulted in four geoelectrical cross-sections indicating the main structural elements (Wannamaker, 1997a) (Figure 1.14). The low resistive zone detected in the Paleozoic sedimentary layer to the southwest of the Sulphur Creek Fault appears to represent (although not necessarily) the hydrothermal aquifer, while the high resistivity area in the upper 500m near wells VC-2B and VC-2A possibly corresponds to a vapor zone.

It is worth mentioning that in spite of the fact that the resistivity anomalies are controlled mainly by faulted structures, the resistivity variations not necessarily reflect the formational boundary or the lithological differences of the rocks mapped in the area. Del Rosario et al. (2005) used the lithology data from rocks penetrated by the nearby wells for geothermal interpretation of the resistivity results. The authors categorized the Malabuyoc geothermal system as a basement aquifer beneath a sedimentary basin with the heated fluid originating east of the survey area. The fluid is channeled along the Middle Diagonal and Montaneza River Faults and emerged along the stretch of the Montaneza River as warm seepages. Figure 1.15 shows the geophysical model of the Malabuyoc thermal system at 250 m below sea level.

Another problem associated with finding the fluid circulation zones by EM methods is that the correlation between low resistivity and fluid concentration is not always correct since alteration minerals produce comparable and often a greater reduction in resistivity. Moreover, although water-dominated

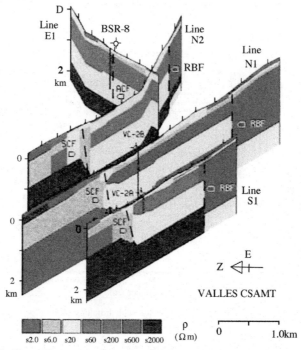

FIGURE 1.14 Fence diagram assembling best-fit 2-D resistivity models for the Sulphur Springs area, USA (after Wannamaker, 1997a). The models are vertically compressed by a factor of two for visibility. Details of the finite-element mesh geometry have been preserved, but resistivities have been grouped into half-decade intervals for simplicity. Surface locations of boreholes are plotted and projected upon the nearest resistivity section. Locations on each section of the Sulphur Creek Fault, Redondo Border Fault, and Alamo Canyon Fault are noted.

geothermal systems have an associated low-resistivity signature, the opposite is not true. So, low-resistivity anomalies are not always suitable as geothermal targets; however, if they are accompanied by low-density anomalies this may increase the probability of such a conclusion. So, integration with geological, other geophysical and drilling data can help to eliminate undesirable low-resistivity targets from consideration (see Chapter 6 below).

Thus, EM sounding of tectonically active areas enables to detect faulted and fractured zones, which, in turn, often indicate the possible location of the hydrothermal circulation.

1.6 MONITORING OF THE TARGET MACROPARAMETERS

Phase change of pore fluid (boiling/condensing) in fractured rocks can result in resistivity changes that are more than one order of magnitude greater than those measured in intact rocks. Secondly, production-induced changes in resistivity can provide valuable insights into the evolution of the host rock and resident

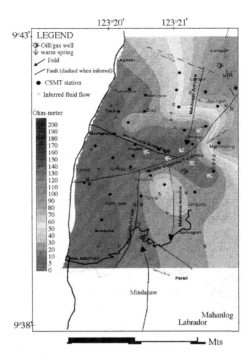

FIGURE 1.15 Geophysical model of the Malabuyoc thermal system, Philippines, at 250 m below sea level (after Del Rosario et al., 2005).

fluids. Therefore, EM monitoring of the faults' macroparameters may provide important information about their activity and fluid content.

Spichak and Popova (2000) developed an artificial neural network expert system for the interpretation of MT monitoring of the fault macroparameters. The authors concluded that the ability of artificial neural network to teach itself by real geophysical (not just EM) data measured at a given location over a sufficiently long period means that the potential exists for using this approach to interpret monitoring data in different geological environments providing that prior knowledge about their resistivity structure is available (in particular, from MT sounding).

The impedance determinant is one of the most suitable data transforms for an adequate interpretation of MT measurements carried out with the aim of monitoring resistivity variations in the geothermal reservoir (Spichak, 2001). The author also showed by means of 3-D numerical modeling that the conductive channel (for instance, a fluid-filled fault or a chain of hot springs) connecting the reservoir with the surface strengthens the effect of the variation in the resistivity inside the reservoir. First, it increases the diameter of the zone of reliable monitoring, and second it reduces the period threshold sufficient for the detection of even small variations in electrical resistivity.

At first glance, the construction of a 3-D geoelectrical model of the zone under study and also the monitoring of crucial parameters would seem to require synchronous measurements carried out at MT sites regularly distributed over the earth's surface. However, forward numerical modeling (Spichak, 2001) indicates that restricting the monitoring to just the electrical conductivity alteration inside the local area in the earth (for instance, a geothermal reservoir) may be sufficient for a proper interpretation of the data regularly measured even at one site at the surface. Since a little 3-D interpretation experience is available in the world, only the above-mentioned computer simulation of both the MT fields' behavior and the monitoring process itself may provide an appropriate basis for effective 3-D MT imaging and monitoring of geothermal areas.

The correct interpretation of the MT monitoring results depends to a great extent on the knowledge about the 3-D resistivity structure of the zone considered. Many seismic and EM studies are carried out aimed at determining "final" three-dimensional models of the regions. It may be that the best way to achieve this goal might be to carry out synchronous measurements of MT data at the sites regularly distributed over the surface. However, such an experiment would result at best in a construction of quite a good 3-D resistivity model that might change unpredictably before the next monitoring measurements are carried out. Thus, the interpretation of the results based on an already inadequate model would lead to an inaccurate estimation of the parameters monitored.

Thus, it is important to use a methodology that will closely link the monitoring results with a knowledge about the internal structure at the moment of the last measurements. In other words, it should enable, the 3-D resistivity model of the studied zone to be updated in accordance with the new MT data collected at its surface or surrounding area. In addition, one should take into account the current 3-D resistivity structure of this zone when interpreting the monitoring results.

Another problem concerns the methodology of the MT survey. The positions of sites as well as the survey regime depend on the aim of the MT sounding. Meanwhile, it would be very tempting to be able to use the results of any MT measurements carried out over the surface both to improve one's knowledge of the 3-D interior and for monitoring purposes. Thus, the problems of effective MT surveying, imaging, and monitoring geothermal zones are threefold and need solving in close relation to each other.

1.7 USING OF GEOLOGICAL AND OTHER GEOPHYSICAL DATA

In order to reduce the uncertainty of EM inversion results (especially when using 1-D or 2-D inversion schemes), it is often useful to add information that comes from geological or other geophysical methods. This can be accomplished in different ways, the most common being an interpretation of the resistivity distribution taking into account geophysical constraints in terms of the temperature well records and gravity or seismic velocity maps (Galanopoulus et al., 1991;

Takasugi et al., 1992; Bibby et al., 1995; Correia and Jones, 1997; Gunderson et al., 2000; Talebi et al., 2005; Ushijima et al., 2005; Harinarayana et al., 2006; Spichak et al., 2007b; Spichak et al., 2013).

In addition to individual DC soundings, Bibby et al. (1995) used over 7000 gravity measurements, temperature and seismic refraction data. Gravity measurements indicated that to a depth of about 2.5 km the upper layers of the Taupo volcanic zone consist of low-density pyroclastic infill. A seismic refraction interface with a velocity change from 3.2 to 5.5 km/s occurs at a similar depth. The cross-sectional area of the convection plumes (identified by DC sounding) appears to increase at depths of 1–2 km, consistent with a decrease in permeability at the depth at which the velocity and density increase. The authors found that the pattern of low-resistivity anomalies detected by DC sounding data delineates the horizontal extent of the near-surface distribution of hot water. Its boundary correlates with the transition from high to low temperatures measured in the drillholes (Figure 1.16).

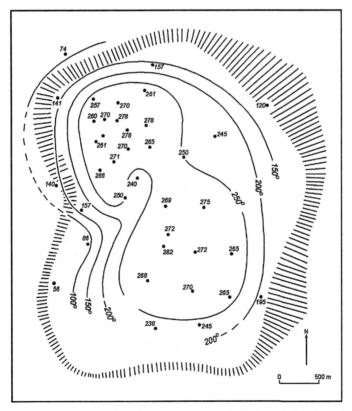

FIGURE 1.16 Correlation of the boundary zone of Broadlands Geothermal Field, determined from resistivity measurements (hatched zone), with temperatures measured in drillholes in the geothermal field (after Bibby et al., 1995). Temperature data are seldom available from outside the boundary zone.

The relationship between electrical resistivity and temperature was the main motivation for the MT survey on the Alentejo Geothermal Anomaly, detected in Hercynian structures in southern Portugal in the area of the intersection of the Messejana fault and the Ferreira–Ficalho overthrust (Correia and Jones, 1997). The results of the survey are discussed in conjunction with magnetic, gravimetric, and seismic information as well as calculations of synthetic temperature distribution within the crust. The area studied appears to be a low electrical resistivity region with a number of deeply rooted high electrical resistivity blocks. The authors concluded that Alentejo Geothermal Anomaly is probably a shallow feature and does not represent the regional thermal regime of the crust in southern Portugal.

Finally, since the interface between the conductive cap (clay alteration halo) and the relatively resistive reservoir is largely controlled by the smectite distribution over the geothermal reservoir, Gunderson et al. (2000) suggest refining the results of MT inversion by measuring the smectite contents in the wells.

A number of joint geological/geophysical data analysis/interpretation techniques are presently available (see, for instance, a review paper by Bedrosian (2007) and references therein). Application of these techniques to the EM and other data measured in the geothermal areas may increase the accuracy of the inversion schemes and result in more robust detection of the reservoir's macroparameters. In particular, Gallardo and Meju (2003) developed a robust 2-D joint inversion scheme incorporating the concept of cross-gradients of electrical resistivity and seismic velocity as constraints so as to investigate the resistivity–velocity relationships in complex near-surface environments more precisely. The results of their joint inversion of DC resistivity and seismic refraction data (Figure 1.17) for the area consisting of a highly fractured granodiorite

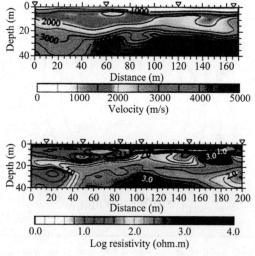

FIGURE 1.17   Optimal 2-D joint velocity (top) and resistivity (bottom) models of the area composed from highly fractured granodiorite rock-mass (after Gallardo and Meju, 2003).

rock-mass suggest that it is possible to distinguish between different types or facies of unconsolidated and consolidated materials. This refines a previously proposed resistivity–velocity interrelationship derived from separate inversions of the respective data sets.

Another approach based on the postinversion joint analysis of different properties was proposed by Spichak et al. (2006). The authors use to this end the method of maximal correlation similitude, which enables to find spatial clasters characterized by maximal correlation between different petrophysical parameters. It was recently used to construct the 3-D resistivity model of the Elbrus volcanic centre (northern Caucasus) based on MT sounding data and maps of tectonic rock fragmentation that were revealed by the lineament pattern analysis (Spichak et al., 2007a).

It is important to be able to include the geophysical constraints in the interpretation procedure so as to avoid the unnecessary "personal" effect. Spichak et al. (1999) proposed the 3-D inversion technique based on Bayesian statistics and Markov chains. The prior geological and geophysical information available is incorporated in the inversion procedure via the probability density function, specified for the prior resistivity palette in the region being investigated, while the parameters to be found are the posterior resistivity values. Posterior parameter uncertainties are also obtained, which are related to the amount and quality of both the input data and prior information.

1.8 CONSTRAINING LOCATIONS FOR DRILLING BOREHOLES

There are two points of view regarding the interpretation of the low-resistivity zone defined in the geothermal area. One is that the low resistivity is a manifestation of the geothermal reservoir itself, which contains hot water in which there are dissolved chemicals. The other is that the low-resistivity zone is regarded as a clay alteration halo overlying the high temperature reservoir. Accordingly, in the former, drilling targets are the low-resistivity objects, whereas in the latter they are the moderately resistive layers underlying the very conductive ones. Therefore, the resistivity structure of the geothermal zone revealed by one of the EM methods should be considered as necessary but not sufficient information for making decisions regarding the drilling targets.

In the presence of the prior geological information about the type of the geothermal system it is often possible to interpret the EM data in geothermal terms and accordingly to determine the drilling targets. In particular, Akiyoshi and Tagomori (2000) interpreted the geoelectrical structure of the Hatchobaru geothermal field taking into consideration that most of geothermal reservoirs in the volcanic areas are of the fractured type and are controlled by the fault system. The authors used the mise-a-la-masse method, as well as MT and controlled-source audio-magnetotelluric, and were able to define the most promising zone for drilling, defined as the low-resistivity areas where electrical discontinuities

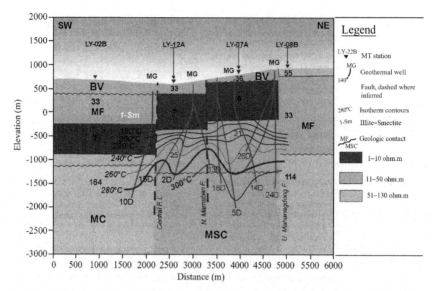

FIGURE 1.18   Resistivity model along profile S-01 in the Mahanagdong geothermal field, Leite, Philippines (after Los Banos and Maneja, 2005).

could be defined. Similarly, Mulyadi (2000) suggests that for future drilling permeable geological structures, such as faults in the resistive areas (45–75 Ωm) should be considered.

However, in the general case it is more reliable to supplement the resistivity data at least by the information from the wells. In particular, the delineated resistivity model (Los Banos and Maneja (2005) shown in Figure 1.18 was correlated with well data such as clay alteration patterns and measured temperatures of the subsurface. The highly conductive middle layer beneath LY-12A and LY-07A coincides with smectite and smectite–illite alteration zones, as well as the 180°C isotherm contour. On the other hand, the underlying moderately resistive layer lies within the illite, biotite, and chlorite alteration zone, which is characterized by temperatures higher than 180°C. Most of the production wells in this sector were drilled within this layer.

Correlation between the low-resistivity and high-temperature zones leads to the conclusion that the influence of clay minerals resulting from hydrothermal alteration processes is not the main contributing factor to enhanced deep conductivities (Romo et al., 1997). The spatial distribution of the deep conductivity anomalies and of the productive wells in the Ahuachapan geothermal area (Figure 1.19) suggests an interesting relationship: in 86% of the cases there is a positive correlation between the presence of a deep conductor and the occurrence of a productive well. This implies the existence of enhanced permeabilities, possibly the result of faulting and fracturing. This correlation, together with the temperature–conductivity relationship, suggests that the low resistivities found can be explained by the combined effect of high temperature and high permeability.

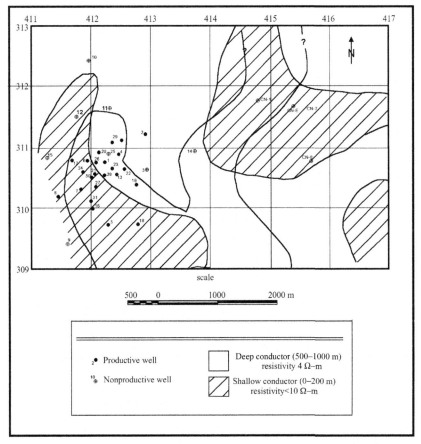

FIGURE 1.19   Deep conductors and drillholes in the Ahuachapan-Chipilapa area (after Romo et al., 1997).

Therefore, decisions regarding drilling target depend on many factors (the main ones being information on the temperature, resistivity, permeability, and alteration mineral distributions) and should be based on all the geological and geophysical information available in the geothermal area.

1.9 CONCLUSIONS

Thus, EM sounding of the geothermal zones enables

- to reveal stratigraphical layering;
- to produce a static image of the reservoir and surrounding structure;
- to locate fractures and faulted zones and determine the strike orientation;
- to detect the boundary between the cap formed by clay minerals and high temperature reservoirs;

- to limit the uncertainties in the distribution of heat flow in the uppermost crust;
- to estimate the permeability values;
- to monitor the phase change of pore fluid in fractured rocks and resident fluids resulting in resistivity changes in the host rocks;
- to improve the accuracy of reservoir temperature estimates;
- finally, to reduce exploration costs.

EM sounding geothermal areas provides a useful contribution to geothermal exploration and exploitation through careful data acquisition, processing, modeling, and interpretation. To take complete advantage of the potential of EM sounding of geothermal zones and distant monitoring macroparameters of the reservoirs, fluid-filled faults and other elements of the geothermal system, it is important to use modern 3-D modeling and inversion techniques, EM data interpretation quantitatively taking into account prior geological information and expert estimates. Integration of EM data with rock physics data, lithology, temperatures, permeability, geological, and other geophysical information may improve the imaging of static and dynamic processes of geothermal systems.

REFERENCES

Akiyoshi, M., Tagomori, K., 2000. Geoelectrical structure of the Hatchobaru geothermal field and the adaptability of electrical methods for its understandings (Expanded Abstr). World Geothermal Congress, Kyushu-Tohoku, Japan.

Ander, M.A., Gross, R., Strangway, D.W., 1984. A detailed magnetotelluric/audiomagnetotelluric study of the Jemez Volcanic Zone, New Mexico. J. Geophys. Res. 89 (B5), 3335–3353.

Anderson, E., Crosby, D., Ussher, G., 2000. Bulls-eye! – Simple resistivity imaging to reliably locate the geothermal reservoir (Expanded Abstr). World Geothermal Congress, Kyushu-Tohoku, Japan, pp. 909–914.

Anderson, E., Jacobo, E., Ussher, G., 1995. A geothermal reservoir revealed – magnetotelluric and data management techniques in a potent combination (Expanded Abstr). World Geothermal Congress, Florence, Italy, pp. 899–902.

Archie, G.E., 1942. The electrical resistivity log as an aid in determining some reservoir characteristics. Trans. AIME 146, 54–67.

Árnason, K., Flóvenz, Ó.G., 1995. Geothermal exploration by TEM-soundings in central Asal Rift in Djibuti, east Africa (Expanded Abstr). World Geothermal Congress, Florence, Italy, pp. 933–938.

Árnason, K., Haraldsson, G.I., Johnsen, G.V., Thorbergsson, G., Hersir, G.P., Saemundsson, K., Georgsson, L.S., Snorrason, S.P., 1986. Nesjavellir-Olkelduhalsr: a geological and geophysical survey 1986. Orkustofnun report OS-86018/JHD-02, 112 pp. (in Icelandic).

Árnason, K., Karlsdottir, R., Eysteinsson, H., Flóvenz, O.G., Gudlaugsson, S.T., 2000. The resistivity structure of high-temperature geothermal systems in Iceland (Expanded Abstr). World Geothermal Congress, Kyushu-Tohoku, Japan, pp. 923–928.

Asaue, H., Koike, K., Yoshinaga, T., Takakura, S., 2006. Magnetotelluric resistivity modeling for 3-D characterization of geothermal reservoirs in the Western side of Mt. Aso, SW Japan. J. Appl. Geophys. 58 (4), 296–312.

Bai, D.H., Meju, M.A., Liao, Z.J., 2001. Magnetotelluric images of deep crustal structure of the Rehai geothermal field near Tengchong, southern China. Geophys. J. Int. 147, 677–687.

Bartel, L.C., Jacobson, R.D., 1987. Results of a controlled-source audio-frequency magnetotelluric survey at the Puhimau thermal area, Kilauea Volcano, Hawaii. Geophysics 52 (5), 665–677.

Bedrosian, P.A., 2007. MT+, integrating magnetotellurics to determine earth structure, physical state and processes. Surv. Geophys. 28, 121–167.

Berktold, A., 1983. Electromagnetic studies in geothermal regions. Geophys. Surv. 6, 173–200.

Bibby, H.M., Caldwell, T., Davey, F.G., Webb, T.H., 1995. Geophysical evidence on the structure of the Taupo Volcanic Zone and its hydrothermal circulation. J. Volcanol. Geotherm. Res. 68, 29–58.

Bromley, C., 1993. Tensor CSAMT study of the fault zone between Waikite and Te Kopia geothermal fields. J. Geomag. Geoelectr. 45, 887–896.

Caglar, I., Tuncer, V., Kaypak, B., Avsar, U., 2005. A high conductive zone associated with a possible geothermal activity around Afyon, Northern part of Tauride zone, South West Anatolia (Expanded Abstr). World Geothermal Congress, Antalya, Turkey.

Correia, A., Jones, F.W., 1997. On the existence of the geothermal anomaly in southern Portugal. Tectonophysics 271, 123–134.

Correia, A., Safanda, J., 2002. Geothermal modeling along a two-dimensional crustal profile in southern Portugal. J. Geodynamics 34, 47–61.

De Lugao, P.P., La Terra, E.F., Kriegshauser, B., Fontes, S.L., 2002. Magnetotelluric studies of the Caldas Novas geothermal reservoir, Brazil. J. Appl. Geophys. 49, 33–46.

Del Rosario, Jr., R.A., Pastor, M.S., Malapitan, R.T., 2005. Controlled source magnetotelluric (CSMT) survey of Malabuyoc thermal prospect, Malabuyoc/Alegria, Cebu, Philippines (Expanded Abstr). World Geothermal Congress, Antalya, Turkey.

Demissie, Y., 2005. Transient electromagnetic resistivity survey at the geysir geothermal field south Iceland (Expanded Abstr). World Geothermal Congress, Antalya, Turkey.

Eysteinsson, H., Hermance, J.F., 1985. Magnetotelluric measurements across the eastern neovolcanic zone in south Iceland. J. Geophys. Res. 90, 10093–10103.

Flóvenz, O.G., Georgsson, L.S., Árnason, K., 1985. Resistivity structure of the upper crust in Iceland. J. Geophys. Res. 90, 10136–10150.

Flóvenz, O.G., Karlsdottir, R., 2000. TEM-resistivity image of a geothermal field in Iceland and the relation of the resistivity with lithology and temperature (Expanded Abstr). World Geothermal Congress, Kyushu-Tohoku, Japan, pp. 1127–1132.

Flóvenz, O.G., Spangerberg, E., Kulenkampff, J., Árnason, K., Karlsdottir, R., Huenges, E., 2005. The role of electrical interface conduction in geothermal exploration (Expanded Abstr). World Geothermal Congress, Antalya, Turkey.

Fulignati, P., Malfitano, G., Sbrana, A., 1997. The Pantelleria caldera geothermal system; data from the hydrothermal minerals. J. Volcanol. Geotherm. Res. 75, 251–270.

Galanopoulus, D., Hutton, V.R.S., Dawes, G.J.K., 1991. The Milos geothermal field: modeling and interpretation of electromagnetic induction studies. Phys. Earth Planet. Inter. 66, 76–91.

Gallardo, L.A., Meju, M.A., 2003. Characterization of heterogeneous near-surface materials by joint 2D inversion of DC resistivity and seismic data. Geophys. Res. Lett. 30 (13), 1658.

Gianelli, G., Mekuria, N., Battaglia, S., Chersicla, A., Garofalo, P., Ruggieri, G., Manganelli, M., Gebregziabher, Z., 1998. Water-rock interaction and hydrothermal mineral equilibria in the Tendaho geothermal system. J. Volcanol. Geotherm. Res. 86, 253–276.

Gonzalez-Partida, E., Garcia Gutierrez, A., Torres Rodriguez, V., 1997. Thermal and petrologic study of the CH-A well from the Chipilapa-Ahuachapan geothermal area, El Salvador. Goethermics 26, 701–713.

Gonzalez-Partida, E., Birkle, P., Torres-Alavarado, I.S., 2000. Evolution of the hydrothermal system at Los Azufres, Mexico, based on petrologic, fluid inclusion and isotopic data. J. Volcanol. Geotherm. Res. 104, 277–296.

Gunderson, R., Cimiming, W., Astra, U., Harvey, C., 2000. Analysis of smectite clays in geothermal drill cuttings by the methylene blue method: for well site geothermometry and resistivity sounding correlation (Expanded Abstr). World Geothermal Congress, Kyushu-Tohoku, Japan, pp. 1175–1181.

Haak, V., Ritter, O., Ritter, P., 1989. Mapping the geothermal anomaly on the island of Milos by magnetotellurics. Geothermics 18 (4), 533–546.

Hafizi, M.K., Aiobi, M., Rahimi, A., 2002. The combination of 2-D and 1-D inversion for 2.5D interpretation of magnetotelluric geothermal sites (Expanded Abstr). EAGE 64th Conference, Florence, Italy.

Harinarayana, T., Abdul Azeez, K.K., Murthy, D.N., Veeraswamy, K., Eknath Rao, S.P., Manoj, C., Naganjaneyulu, K., 2006. Exploration of geothermal structure in Puga geothermal field, Ladakh Himalayas, India, by magnetotelluric studies. J. Appl. Geophys. 58 (4), 280–295.

Hermance, J.F., 1985. Magnetotelluric measurements across the Eastern neovolcanic zone in south Iceland. J. Geophys. Res. 90 (B12), 10093–10103.

Hoover, D.B., Frischknecht, F.C., Tippens, C., 1976. Audiomagnetotelluric soundings as reconnaissance exploration technique in Long Valley, California. J. Geophys. Res. 81, 801–809.

Hoover, D.B., Long, C.L., Senterfit, R.M., 1978. Some results from audiomagnetotelluric investigations in geothermal areas. Geophysics 43, 1501–1514.

Ingham, M.R., 1991. Electrical conductivity structure of the Broadlands-Ohaaki geothermal field, New Zealand. Phys. Earth Planet. Inter. 66, 62–75.

Kajiwara, T., Mogi, T., Fomenko, E., Ehara, S., 2000. Three-dimensional modeling of geoelectrical structure based on MT and TDEM data in Mori geothermal held, Hokkaido, Japan (Expanded Abstr). World Geothermal Congress, Kyushu-Tohoku, Japan, pp. 1313–1318.

Lackschewitz, K.S., Singer, A., Botz, R., Schonberg, G.D., Stoffers, P., Horz, K., 2000. Formation and transformation of clay minerals in the hydrothermal deposits of Middle Valley, Juan de Fuca Ridge, ODP Leg 169. Econ. Geol. 95, 361–389.

Lagios, E., Galanopoulos, D., Hobbs, B.A., Dawes, G.J.K., 1998. Two-dimensional magnetotelluric modelling of the Kos Island geothermal region (Greece). Tectonophysics 287, 157–172.

Layugan, D.B., Rigor, Jr., D.M., Apuada, N.A., Los Bafios, C.F., Olivar, R.E.R., 2005. Magnetotelluric (MT) resistivity surveys in vanous geothermal systems in central Philippines (Expanded Abstr). World Geothermal Congress, Antalya, Turkey.

Long, C.L., Kaufman, H.E., 1980. Reconnaissance geophysics of a known geothermal resource area, Weiser, Idaho, and Vale, Oregon. Geophysics 45, 312–322.

Los Banos, C.F., Maneja, F.C., 2005. The resistivity structure of the Mahanagdong geothermal field, Leyte, Philippines (Expanded Abstr). World Geothermal Congress, Antalya, Turkey.

Manzella, A., Spichak, V., Pushkarev, P., Sileva, D., Oskooi, B., Ruggieri, G., Sizov, Yu., 2006. Deep fluid circulation in the Travale geothermal area and its relation with tectonic structure investigated by a magnetotelluric survey (Expanded Abstr). 31th Workshop on Geothennal Reservoir Engineering. Stanford University, Stanford, USA.

Meju, M.A., 2002. Geoelectromagnetic exploration for natural resources: models, case studies and challenges. Surv. Geophys. 23, 133–205.

Mogi, T., Nakama, S., 1993. Magnetotelluric interpretation of the geothermal system of the Kuju volcano, southwest Japan. J. Volcanol. Geotherm. Res. 56, 297–308.

Mulyadi, 2000. Magnetic Telluric method applied for geothermal exploration in Sibayak. North Sumatra (Proceedings Expanded Abstr). Geothermal Congress, Kyushu-Tohoku, Japan, pp. 1469–1472.

Munõz, G., 2014. Exploring for geothermal resources with electromagnetic methods. Surv Geophys. 35, 101–122 DOI 10.1007/s10712-013-9236-0.

Natland, J.-H., Dick, H.J.B., 2001. Formation of the lower ocean crust and the crystallization of gabbroic cumulates at a very slowly spreading ridge. J. Volcanol. Geotherm. Res. 110, 191–233.

Okada, H., Yasuda, Y., Yagi, M., Kai, K., 2000. Geology and fluid chemistry of the Fushime geothermal field, Kyushu, Japan. Geothermics 29, 279–311.

Oskooi, B., Pedersen, L.B., Smirnov, M., Árnason, K., Eysteinsson, H., Manzella, A., 2005. The deep geothermal structure of the Mid-Atlantic Ridge deduced from MT data in SW Iceland. Phys. Earth Planet. Int. 150, 183–195.

Patrier, P., Papapanagiotou, P., Beaufort, D., Traineau, H., Bril, H., Rojas, J., 1996. Role of permeability versus temperature in the distribution of the fine (< 0.2 mu m) clay fraction in the Chipilapa geothermal system (El Salvador, Central America). J. Volcanol. Geotherm. Res. 72, 101–120.

Pellerin, L., Johnston, J.M., Hohmann, G.W., 1996. A numerical evaluation of electromagnetic methods in geothermal exploration. Geophysics 61, 121–130.

Pérez-Flores, M.A., Antonio-Carpio, R.G., Gomez-Treviño, E., 2005. Tridimensional inversion of DC resistivity data (Expanded Abstr). World Geothermal Congress, Antalya, Turkey.

Pérez-Flores, M.A., Gomez-Treviño, E., 1997. Dipole–dipole resistivity imaging of the Ahuachapan–Chipilapa geothermal field, El Salvador. Geothermics 26 (5/6), 657–680.

Pérez-Flores, M.A., Schultz, A., 2002. Application of 2-D inversion with genetic algorithms to magnetotelluric data from geothermal areas. Earth Planets Space 54, 607–616.

Risk, G.F., Bibby, H.M., Bromley, C.J., Caldwell, T.G., Bennie, S.L., 2002. Appraisal of the Tokaanu-Waihi geothermal field and its relationship with the Tongariro geothermal field, New Zealand. Geothermics 31, 45–68.

Romo, J.M., Flores, C., Vega, R., Vázquez, R., Pérez Flores, M.A., Treviño, E.G., Esparza, F.J., Quijano, J.E., García, V.H., 1997. A closely-spaced magnetotelluric study of the Ahuachapan-Chipilapa geothermal field, El Salvador. Geothermics 26, 627–656.

Romo, J.M., Wong, V., Flores, C., 2000. The subsurface electrical conductivity and the attenuation of coda waves at Las Tres Virgines geothermal field in Baja California Sur, Mexico (Expanded Abstr). World Geothermal Congress, Kyushu-Tohoku, Japan, pp. 1645–1650.

Sener, M., Gervek, A.I., 2000. Distribution and significance of hydrothermal alteration minerals in the Tuzla hydrothermal system, Canakkale, Turkey. J. Volcanol. Geotherm. Res. 96, 215–228.

Shen, J., Sasaki, Y., Ushijima, K., 2000. Reconstruction of resistivity and permittivity profiles using electromagnetic well logging data (Expanded Abstr). World Geothermal Congress, Kyushu-Tohoku, pp. 1731–1736.

Shmonov, V.M., Vitovtova, V.M., Zharikov, A.V., 2000. Fluid permeability of the earth crust rocks (in Russian). Scientific World, Moscow, p. 276.

Spichak, V.V., 2001. Three-dimensional interpretation of MT data in volcanic environments (computer simulation). Ann. Geofis. 44 (2), 273–286.

Spichak, V.V., 2002. Advanced three-dimensional interpretation technologies applied to the MT data in the Minamikayabe thermal area (Hokkaido, Japan) (Expanded Abstr). EAGE 64th Conference, Florence, Italy.

Spichak, V.V., Borisova, V.P., Fainberg, E.B., Khalezov, A.A., Goidina, A.G., 2007a. Electromagnetic 3-D tomography of the Elbrus volcanic center according to magnetotelluric and satellite data. J. Volcanol. Seismol. 1 (1), 53–66.

Spichak, V.V., Manzella, A., 2009. Electromagnetic sounding of geothermal zones. J. Appl. Geophys. 68, 459–478.

Spichak, V.V., Menvielle, M., Roussignol, M., 1999. Three-dimensional inversion of MT data using Bayesian statistics. In: Spies, B., Oristaglio, M. (Eds.), 3-D electromagnetics, SEG monograph. GD7, Tulsa, USA, pp. 406–417.

Spichak, V.V., Popova, I.V., 2000. Artificial neural network inversion of MT – data in terms of 3-D earth macroparameters. Geophys. J. Int. 42, 15–26.

Spichak, V., Rybin, A., Batalev, V., Sizov, Yu., Zakharova, O., Goidina, A. 2006. Application of ANN techniques to combined analysis of magnetotelluric and other geophysical data in the northern Tien Shan crustal area (Expanded Abstr). 18th IAGA WG 1.2 Workshop on Electromagnetic Induction in the Earth, El Vendrell, Spain.

Spichak, V.V., Schwartz, Ya., Nurmukhamedov, A., 2007b. Conceptual model of the Mutnovsky geothermal deposit (Kamchatka) based on electromagnetic, gravity and magnetic data (Expanded Abstr). EAGE Workshop on "Innovation in EM, Gravity and Magnetic methods: a new perspective for exploration," Capri, Italy.

Spichak, V.V., Zakharova, O.K., Goidina, A.G., 2013. A new conceptual model of the Icelandic crust in the Hengill geothermal area based on the indirect electromagnetic geothermometry. J. Vol anol. Geotherm. Res. 257, 99–112.

Srodon, J., 1999. Nature of mixed-layer clays and mechanisms of their formation and alteration. Annu. Rev. Earth Planet. Sci. 27, 19–53.

Suharno, S., Browne, P.R.L., Soengkono, S., Prijanto, Sudarman, S., 2000. A geophysical model and the subsurface geology at the Ulubelu geothermal area, Lampung, Indonesia (Abstract). AAPG Bull. 84, 1498.

Svetov, B., 2006. Seismoelectrical methods of Earth study. In: Spichak, V.V. (Ed.), Electromagnetic sounding of the Earth's interior. Elsevier, Amsterdam, pp. 79–101.

Takasugi, S., Tanaka, K., Kawakami, N., Muramatsu, S., 1992. High spatial resolution of the resistivity structure revealed by a dense network MT measurement – a case study in the Minamikayabe Area, Hokkaido, Japan. Geomagn. Geoelectr. 44, 289–308.

Talebi, B., Khosrawi, K., Ussher, G., 2005. Review of resistivity surveys from the NW Sabalan geothermal field, Iran (Expanded Abstr). World Geothermal Congress, Antalia, Turkey.

Uchida, T., 1995. Resistivity structure of Sumikawa geothermal field, northeastern Japan, obtained from magnetotelluric data (Expanded Abstr). World Geothermal Congress, Florence, Italy, pp. 921–925.

Uchida, T., 2005. Three-dimensional magnetotelluric investigation in geothermal fields in Japan and Indonesia (Expanded Abstr). World Geothermal Congress, Antalia, Turkey.

Uchida, T., Ogawa, Y., Takakura, S., Mitsuhata, Y., 2000. Geoelectrical investigation of the Kakkonda geothermal field, northern Japan (Expanded Abstr). World Geothermal Congress, Kyushu-Tohoku, Japan, pp. 1893–1898.

Uchida, T., Song, Y., Lee, T.J., Mitsuhata, Y., Lim, S.K., Lee, S.K., 2005. Magnetotelluric survey in an extremely noisy environment at the Pohang low-enthalpy geothermal area, Korea (Expanded Abstr). World Geothermal Congress, Antalia, Turkey.

Ushijima, K., Mustopa, E.J., Jotaki, H., Mizunaga, H., 2005. Magnetotelluric soundings in the Takigami geothermal Aria, Japan (Expanded Abstr). World Geothermal Congress, Antalia, Turkey.

Ushijima, K., Noritomi, K., Tagomori, K., Kinoshita, Y., 1986. Joint inversion of MT and DC resistivity data at Hatchobaru area. Geotherm. Resour. Council Trans. 10, 243–246.

Ussher, G., Harvey, C., Johnstone, R., Anderson, E., 2000. Understanding the resistivities observed in geothermal systems (Expanded Abstr). World Geothermal Congress, Kyushu-Tohoku, Japan, pp. 1915–1920.

Veeraswamy, K., Harinarayana, T., 2006. Electrical signatures due to thermal anomalies along mobile belts reactivated by the trail and outburst of mantle plume: evidences from the Indian subcontinent. J. Appl. Geophys. 58 (4), 313–320.

Volpi, G., Manzella, A., Fiordelisi, A., 2003. Investigation of geothermal structures by magnetotellurics (MT): an example from the Mt. Amiata area, Italy. Geothermics 32, 131–145.

Wannamaker, P.E., 1997a. Tensor CSAMT survey over the Sulphur Springs thermal area, Valles Caldera, New Mexico, U.S.A. Part I: Implications for structure of the western Caldera. Geophysics 62, 451–465.

Wannamaker, P.E., 1997b. Tensor CSAMT survey over the Sulphur Springs thermal area, Valles Caldera, New Mexico, U.S.A. Part II: Implications for CSAMT methodology. Geophysics 62, 466–476.

Wannamaker, P.E., Jiracek, G.R., Stodt, J.A., Caldwell, T.G., Gonzales, V.M., McKnight, J.D., Porter, A.D., 2002. Fluid generation and pathways beneath an active compressional orogen, the New Zealand southern Alps, inferred from magnetotelluric data. J. Geophys. Res. 107 (6), 6-1–6-22.

Wannamaker, P.E., Rose, P.E., Doemer, W.M., McCulloch, J., Nurse, K., 2005. Magnetotelluric surveying and monitoring at the Coso Geothermal Area, California, in support of the enhanced geothermal systems concept: survey parameters, initial results (Expanded Abstr). World Geothermal Congress, Antalia, Turkey.

Ward, S.H., 1990. Resistivity and induced polarization methods. In: Ward, S.H. (Ed.), Geotechnical and environmental geophysics, vol. 1 – Review and tutorial. Soc. Explor. Geophys, pp. 147–189.

Waxman, M.H., Smith, J.M., 1968. Electrical conductivities in oil-bearing Shaley sands. Soc. Pet. Eng. J., 102–122, SPE Paper 1863-A at SPE Ann. Fall Meeting, Houston.

Yang, K., Browne, P.R.L., Huntington, J.F., Walshe, J.L., 2001. Characterising the hydrothermal alteration of the Broadlands-Ohaaki geothermal system, New Zealand, using short-wave infrared spectroscopy. J. Volcanol. Geotherm. Res. 106, 53–65.

Yang, K., Huntington, J.F., Browne, P.R.L., Ma, C., 2000. An infrared spectral reflectance study of hydrothermal alteration minerals from the Te Mihi sector of the Wairakei geothermal system, New Zealand. Geothermics 29, 377–392.

Zhang, S., Paterson, M.S., Cox, S.F., 1994. Porosity and permeability evolution during hot isostatic pressing of calcite aggregates. J. Geophys. Res. 99, 15741–15760.

Zlotnicki, J., Vargemezis, G., Mille, A., Bruere, F., Hammouya, G., 2006. State of the hydrothermal activity of Soufriere of Guadeloupe volcano inferred by VLF surveys. J. Appl. Geophys. 58 (4), 265–279.

Chapter 2

Techniques Used for Estimating the Temperature of the Earth's Interior

Chapter Outline

Estimation of deep temperatures is one of the key factors in both the study of the geothermal processes in the earth's interior (see Clauser (2009) and references therein) and in solving of different tasks of applied geothermics (see Eppelbaum et al. (2014) and references therein). In this chapter we will briefly outline the approaches used to this end and finally demonstrate an example of building of the temperature model of the mantle from the magnetovariational data.

2.1 TEMPERATURE MODELS BASED ON THE BOREHOLES' LOGS AND THE HEAT FLOW DATA

Actual data about the measured temperatures are limited to the borehole depths amounting in most cases to 1–2 km. So, in order to get an idea of spatial temperature distribution in the studied area it is necessary to carry out interpolation/extrapolation of temperature logs measured usually in nonevenly distributed and

Electromagnetic Geothermometry. http://dx.doi.org/10.1016/B978-0-12-802210-8.00002-2
37

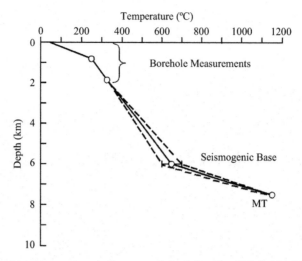

FIGURE 2.1   Linear temperature gradient extrapolation below the well logs (after Foulger, 1995).

often not numerous wells, which often results in considerable errors. Pribnow and Hamza (2000) have estimated the vertical temperature distribution in the Soultz-sous-Forêts (France) geothermal area by lateral averaging of the temperature well logs from all boreholes located in the band of 20 km width from each side of the considered profile (see Figure 7.2 from the Chapter 7). As it was shown by Spichak (2006), temperature interpolation using the geostatistical tools and artificial neural network-based techniques may decrease the relative temperature interpolation errors up to 27–30% and 12–15%, accordingly (see also Chapter 3).

Temperature values beneath the boreholes are often estimated using linear extrapolation of the temperature gradients determined from the heat flow data (Foulger, 1995; Björnsson, 2008) (Figure 2.1). However, this approach is implicitly based on the assumptions that the lithosphere is homogeneous and the heat transfer mechanism is conductive, which rarely takes place in the geothermal areas and may lead to erroneous temperature estimations at given depths (see in this connection also Section 8.5.1 in the Chapter 8).

To cope with the lack of the temperature data, indirect measurements such as the surface heat flow Q_s can be used to model the subsurface temperature according to equation (2.1) for 1-D heat conduction (Limberger and van Wees, 2013):

$$T(z) = \frac{-A}{2k} z^2 + \frac{Q_s + Ah}{k} z + T_0, \qquad (2.1)$$

where the temperature at depth $T(z)$ is related to the layer thickness h, its thermal conductivity k (supposed to be known constant), the radiogenic heat production A and the surface temperature T_0.

Following this approach Limberger and van Wees synthesized the 3-D temperature model for the Europe (Figure 2.2) from 1-D profiles (2.1), with horizontal

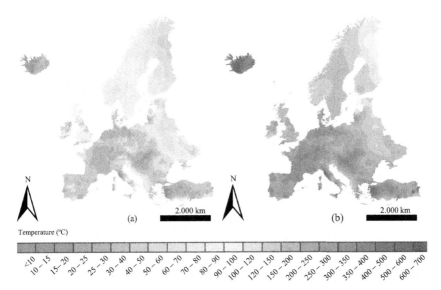

FIGURE 2.2 Depth slices taken from the 3-D temperature model at depths 5 km (a) and 10 km (b) (after Limberger and van Wees, 2013).

resolution of 10×10 km based on the two-layered model reduced from the European crust model (Tesauro et al., 2008) and heat flow data (Artemieva, 2006).

Solution of the heat flow equation in 2-D or 3-D statements is based on an assumption about steady state of the heat flows at lateral boundaries of the modeling domain and on prior knowledge of the heat flow (temperature) at the upper and lower boundaries of the area (Podgornykh et al., 2001; Ollinger et al., 2010). This approach may, however, fail at active margins or in areas where a significant fluid migration is suspected. In such zones the thermal field is obviously non-steady-state, and the calculation based on a stationary model may lead to significant errors.

Methods for estimating the temperatures allowing for geological processes in time permit, in principle, construction of more adequate models (see, for instance Alexidze et al. (1993)). However, this approach requires knowledge on the physical nature of the processes and the geological conditions (lithology, absolute age of the formations, start date and duration of volcanism in a geological past, and so on). Lack of necessary geological and geophysical data makes such a geothermal modeling difficult.

2.2 TEMPERATURE ESTIMATIONS USING INDIRECT GEOTHERMOMETERS

Global studies of hydrothermal processes showed that specific properties of the underground fluid composition are closely related with the geothermal conditions of their formation. Therefore, studying of these properties provides indirect information about the thermal state of the interior that complements the results of direct thermometry and serves as a basis for forecasting the deep geothermal conditions in scantily explored regions.

Established experimentally is a temperature dependency of the composition of some characteristic hydrothermal components. Using empirical or semiempirical formulas, one can estimate the "base depth" temperature from the known amount or proportion of these components in areas of surface manifestations of thermal activity. To this end indirect geothermometers are usually used. Below we briefly discuss them following (Spichak and Zakharova, 2012).

2.2.1 Mineral Geothermometers

Stability of the alteration minerals' assemblages in different temperature ranges is used for indirect temperature estimation at depth. An empirical relationship between formation temperature and the occurrence of specific alteration minerals is used to determine a proper depth for the production casing. This method offers currently the best estimation of aquifer production temperature that can be made during drilling. Therefore, aquifers that are too cold for production can be excluded from the production part of a well based on the absence or presence of certain index minerals.

Mixed-layer clay geothermometers have proved useful in the study of many geothermal systems around the world (see, for instance, Harvey and Browne (1991) and references therein). The earliest investigations were based on the dioctahedral smectite to illite transformation, but it was subsequently extended to other transformations such as trioctahedral smectite to chlorite. More recently, numerous other mixed-layer structures that may have significance as sensitive indicators of changes in temperature have been recognized. Mixed-layer clay geothermometers have proved most effective in sediments or tuff sequences but may not give correct results in fracture-dominated geothermal reservoirs, since incomplete water–rock interaction away from major flow paths may invalidate their use.

2.2.2 Hydrochemical Geothermometers

To hydrochemical geothermometers refer: dissolved silica content, atomic and ion Na/K and Na/Li ratios, and Na, K, and Ca concentration proportions (see Kharaka and Mariner (1989) and references therein). The use of these parameters is based on the assumptions of (i) equilibrium in the water–rock system within the zone of hydrothermal formation and (ii) absence of precipitation–dissolution of the given component along the path of water migration from the heating zone (thermal supply) to the probing point. Readings of these thermometers depend on many factors including temperature, pressure, hydrothermal flow speed, mineralogical conditions, partial pressure of gases, pH of the medium and other.

One of the most widespread chemical geothermometers is a silica one. This is due to the fact that the dissolubility of silica contained in the solution as $Si(OH)_4$ molecules depends strongly on temperature and weakly – on the other ions content within a wide pH range; usually silica is depositing quite slowly.

Estimations calculated from Si-geothermometer forecast the temperature in deepest parts of hydrothermal systems.

Presence of sea water, the water salinity, or the nature of the rocks surrounding the reservoirs can influence the temperature values given by hydrochemical geothermometers. For instance, the silica geothermometer underestimates the reservoir temperature when applied to deep geothermal fluids diluted by surface waters or after silica precipitation due to a fluid cooling. Conversely, for dilute thermal waters collected from volcanic or granite areas, the Na/K geothermometer often yields overestimated reservoir temperatures.

2.2.3 Gas Geothermometers

Gas geothermometers are based on equilibrium chemical reactions between gaseous species. For each reaction considered a thermodynamic equilibrium constant may be written, where the concentration of each species is represented by its partial pressure in vapor phase. The concentrations (or ratios) of gases like CO_2, H_2S, H_2, N_2, NH_3, and CH_4 are controlled by temperature (see, for instance, Arnorsson and Gunnlaugsson (1985) and references therein). Because of that, these data are used to study a correlation between the relative gas concentrations and the temperature of the reservoir.

The gas–gas equilibrium in geothermal fields with two phase components should not reflect the real gas composition present in the reservoir. It depends on many factors like gas/steam ratio. No reequilibration of chemical species from the source or sources to well head is assumed. Application of geothermometric techniques based on thermodynamic equilibrium of organic gases is a reliable tool to evaluate the temperature of deep systems even by adopting the hydrocarbon composition of natural discharges.

2.2.4 Isotopic Geothermometers

In isotopic geothermometry the δD, $\delta^{18}O$, $\delta_{CO_2}^{13}C$, $\delta_{CH_4}^{13}C$, $^3He/^4He$ quantities are used. In particular, studies showed that the ratio $^3He/^4He$ in underground fluids is a stable regional marker; it remains virtually the same within the given region for different type fluids, but differs considerably from region to region, thereby reflecting their tectonic specificity.

Regional $^3He/^4He$ variations in the earth's gases agree with heat flow variations. The established empirical dependency between these parameters (Polyak and Tolstikhin, 1985) allows rough heat flow estimation from the isotopic-helium ratio in a gaseous phase of the source.

2.2.5 Geothermometers Based on the Seismicity Bounds

Finally, it is worth mentioning also an approach often used for the temperature assessment in seismically active areas. The earthquakes' hypocenters trace the depth at which brittle/ductile transition of rocks is still possible. Correspondingly,

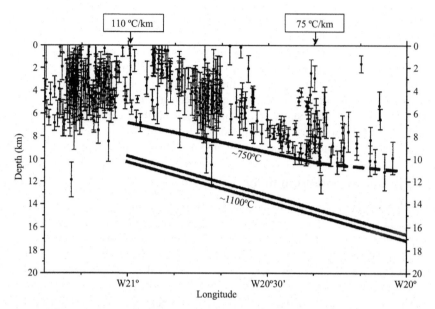

FIGURE 2.3   Location of the basalt solidus ($T = 750°C$) in the Icelandic crust derived from the seismicity bounds (after Björnsson, 2008).

the assumption concerning the rock type and the pressure indirectly defines the maximum temperature in a particular hypocenter or at a particular depth, if the temperature distribution is laterally homogeneous. Tryggvason et al. (2002) and Björnsson (2008) have estimated in this way the depth of the basalt solidus in the Icelandic crust (Figure 2.3).

2.3   INTERPLAY BETWEEN THE ELECTROMAGNETIC SOUNDING DATA, ROCK PHYSICS, AND PRIOR GEOLOGICAL INFORMATION

Estimating the electrical conductivity in the earth from the electromagnetic induction data, one could provide a direct quantitative assessment of the temperature values in the same locations using some empirical formula. Laboratory experiments on electrical conductivity as a function of temperature have shown that it generally increases with increasing temperature. However, the conductivity range and the rate of increase are different for the various types of rocks, since the electrical properties of saturated rocks are also sensitive to such factors as fluid composition, water and clay content, porosity and other microstructural parameters (see, for instance, Roberts (2002) for more details).

In general, the melt composition, the fraction of partial melts and the presence of water greatly varies the resistivity of rocks (e.g., Lebedev and Khitarov, 1964; Wannamaker, 1986). Due to this one may encounter the ambiguous situation where the same resistivity can be related to the different temperature values.

For example, the electrical resistivity of about 2 Ωm, such as the one identified by Frischknecht (1967) in his study on Kilauea Iki lava lake molten basaltic magma or that of Hoffmann-Rothe et al. (2001) below Java can be attributed to the molten magmatic lava at a temperature of above 1250°C, dry rhyolites at 950°C, the melt with 1 wt% H_2O at 875°C, to the melt with 3 wt% H_2O at 830°C, or the aqueous melt with 6 wt% H_2O at 760°C.

The sensitivity of resistivity can be, in principle, used for estimating the temperature if we reduce the multiparametric space of the parameters enumerated above to a single resistivity–temperature hyperplane, for example, by involving the prior geological and/or geophysical data or rock physics. Extensive laboratory studies of the rock samples acquired from different regions were carried out under different physical conditions in order to find a more exact relationship between the resistivity and temperature. Some results of these studies are formulated as empirical dependences (Arps, 1953; Dakhov, 1962; Waff, 1974; Shankland and Waff, 1977; Ucok et al., 1980; Flóvenz et al., 1985; Revil et al., 1998; Gaillard, 2004; Harinarayana et al., 2006) while others are presented in the tabular form (Coster, 1948; Parkhomenko, 1967; Bondarenko, 1968; Duba et al., 1974; Volarovich and Parkhomenko, 1976; Olhoeft and Ucok, 1977; Rai and Manghnani, 1978a, 1978b; Haak, 1980; Caldwell et al., 1986; Llera et al., 1990; Tyburczy and Fisler, 1995; Roberts, 2002; Kristinsdottir et al., 2010; Pommier et al., 2010).

A common way of the temperature estimation based on this approach consists in two steps: (i) determining of the electrical resistivity as a function of coordinates (in particular, depth) from the electromagnetic sounding data; (ii) a conversion of the resistivity values into the temperature ones by comparing them with the results of the laboratory studies of the rock samples collected in the study area (Hermance and Grillot, 1970, 1974; Shankland and Ander, 1983; Ussher et al., 2000; Ledo and Jones, 2005).

For converting the resistivity values derived from the electrical sounding into the temperature ones without study of such samples *in situ*, it is required to have at least a petrological information or guess about the lithology. For example, Hermance and Grillot (1970) estimated the crustal temperature in Iceland at a depth of 15 km assuming that the crust is composed of basalt or peridotite. These authors compared the temperature dependences of electrical conductivity for different rock samples collected elsewhere with the dependence determined at this depth from the results of magnetotelluric sounding (Figure 2.4).

It is worth mentioning in this connection that the rightmost graph of the Figure 2.4 was not included in the analysis in the quoted study because, in the opinion of its authors, the temperature predicted by this graph at a depth of 15 km is unfeasibly low (about 250°C). However, the estimates carried out by Spichak et al. (2013) using another approach argue for these values being quite probable for the Icelandic crust (see Chapter 8 for more details). This example shows that such estimates should be treated with a good deal of caution since the results of the resistivity–temperature conversion strongly depend on the representativeness of the set of the rock samples used to this end.

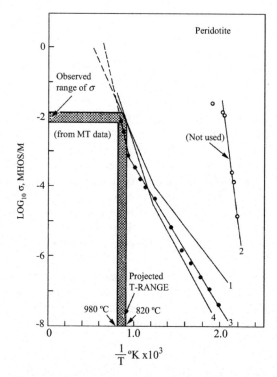

FIGURE 2.4 The laboratory data on the electrical conductivity of peridotite (after Hermance and Grillot, 1974) according to different authors (1, 2, and 4, after Parkhomenko, 1967; 3, after Coster, 1948).

The systematic application of this approach to estimating the temperature revealed a number of challenges. First, the real material properties (i.e., the metamorphism of the rocks) are not known. Second, the different degree of oxidation of the rock samples with close compositions can significantly variegate their resistivity at equal temperatures (Duba et al., 1974). Third, the resistivity not only depends on temperature but is also sensitive to pressure and the latter dependence is often uncertain. At last, the equilibrium between the metamorphism of the rocks and the temperature *in situ* can differ from that studied in the laboratory experiments with the rock samples.

From the mathematical point of view, this is reflected in the fact that neither resistivity nor temperature in the developed empirical formulas depends on the spatial coordinates. Another important factor is that the possibility of deriving such formulas by itself is based on existence of the correlation between the electrical conductivity and temperature over the considered set of the core samples. Therefore, their application in the locations where such a correlation could be disturbed by unknown external factors (like alteration mineralogy, lithology, etc.) may lead to erroneous inferences.

Meanwhile, if one has an idea about the class of the models describing such dependence, it is, in principle, possible to determine its unknown parameters directly from the electromagnetic data by solving the appropriate optimization task. In the next section we will exemplify this statement by building of the mantle temperature model from the magnetovariational data.

2.4 TECHNIQUE FOR THE DEEP TEMPERATURE MODEL BUILDING USING THE GLOBAL MAGNETOVARIATIONAL SOUNDING DATA AND GUESS ABOUT THE CONDUCTANCE MECHANISMS

There are a few approaches to estimating temperature distribution in the earth's deep interior (see, for instance, Artemieva (2006) and references therein). The geo-electrical data helps to interpret deep temperature profiles based on estimated thermal conductivity and heat generation of rocks and on extrapolation of measured temperature gradients and heat flow values (Adam, 1976; Cermak and Lastovicko-va, 1987). In particular, electrical conductivity in the upper mantle constrains such extrapolation as conductivity anomalies are possible only in a limited temperature range. Averaged curves of electrical conductivity as functions of temperature for different types of rocks (granites, basalts, ultrabasics, etc.) as well as guess about the electrical conductance mechanisms were used to estimate its generalized distributions in the regional scale. During last decades a number of 1-D electrical conductivity models for different cratonic regions of the world was built (Rikitake, 1966; Safonov et al., 1976; Stacey, 1977; Vanyan et al., 1977; Vanyan and Cox, 1983; Karato, 1990; Constable et al., 1992; Korja, 1993; Kurtz et al., 1993; Schultz et al., 1993; Mareschal et al., 1995; Singh et al., 1995; Jones, 1999; Neal et al., 2000; Korja et al., 2002; Wu et al., 2002; Jones et al., 2003) (Figure 2.5).

In order to convert the electrical resistivity profiles into the temperature ones without rock physics data it is required to guess the conductance mechanisms. Dmitriev et al. (1975) suggested estimating the mantle distribution of the temperature from the global magnetovariational sounding (MVS) data. (MVS is one of the sounding techniques based on measuring of only the earth's magnetic field – see Berdichevsky et al. (2007) for more details.) Below, we consider this approach following (Dmitriev et al., 1975) and subsequent articles of these authors.

2.4.1 Analysis of the Mantle Temperature Models

To a first approximation, the known temperature distributions in the nonconvective mantle (Figure 2.6) could be analytically described by the following formula:

$$T(r) = T_0 + T_1 \ th\frac{R-r}{\alpha}, \tag{2.2}$$

where R is the earth's radius, r is the radius of the observation point, T_0 (°K) is the earth's surface temperature, $T_0 + T_1$ is temperature in the center of the earth, and α is the parameter characterizing the rate of the depth variations in temperature. This class of models was used as the basis for solving the stated problem by Dmitriev et al. (1975, 1977a, 1977b, 1986, 1988).

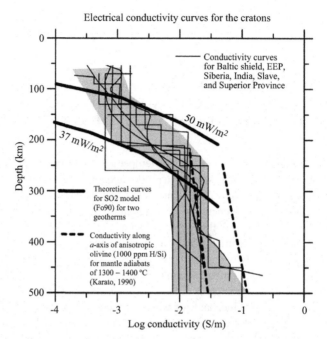

FIGURE 2.5   Electrical conductivity profiles for different cratonic regions of the world (after Artemieva, 2006): Baltic shield (Korja, 1993; Korja et al., 2002), East European craton (Vanyan et al., 1977), Siberia (Safonov et al., 1976; Vanyan and Cox, 1983; Singh et al., 1995), India (Singh et al., 1995), Slave craton (Wu et al., 2002; Jones et al., 2003), and Superior Province (Schultz et al., 1993; Kurtz et al., 1993; Mareschal et al., 1995; Neal et al., 2000). Also shown are synthetic conductivity curves calculated for conductive continental geotherms of 37 and 50 mW/m² for the standard olivine model S02 (Constable et al., 1992), with 90% of forsterite (thick solid lines) and conductivity along a-axis of anisotropic wet olivine (1000 ppm H/Si) for mantle adiabats of 1300–1400°C (Karato, 1990) (thick dashed lines).

The dependence of the electrical conductivity on temperature can be expressed as a sum of three exponential terms, which correspond to the impurity conductance, ionic conductance, and electronic conductance itself:

$$\sigma = \sigma_{0m}\exp(-E_n\,/\,kT) + \sigma_{0i}\exp(-E_i\,/\,kT) + \sigma_{0e}\exp(-E_e\,/\,2kT), \quad (2.3)$$

where E_m, E_i, E_e are the activation energies for the impurity, ionic, and electronic conductances, and k is the Boltzmann constant.

The first two terms describe the conductivity distributions in the earth's crust and upper mantle (down to a depth of ~400 km), respectively. Since it is assumed that below this level the conductance is electronic and the effects at the shallower depths cannot be detected from the magnetovariational data (Dmitriev et al., 1986, 1988), one can build the temperature distribution $T(r)$ in the mantle using only the last term of (2.3). Substituting the relationship (2.2) into (2.3) and making some algebra, we obtain

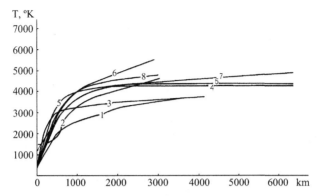

FIGURE 2.6 The temperature distributions reported in the literature (after Dmitriev et al., 1975): 1 – the average temperature, the mantle convection taken into account (Stacey, 1977); 2 – the temperature distribution, the mantle convection taken into account (Tozer, 1959); 3 – the oceanic model without convection (Stacey, 1977); 4 – the distribution according to (Zharkov et al., 1971); 5 – the temperature distribution at a constant heat conduction (Lyubimova, 1968); 6 – the distribution estimated from seismic velocity data (Zharkov, 1958); 7 – the distribution determined by the reference point method (Zharkov et al., 1971); 8 – the distribution based on the electrical conductivity data (Tozer, 1959).

$$\sigma(r) = \sigma_0(r) + \gamma_1 \exp\left[-\frac{E}{2kT_1} \frac{1 - th\dfrac{R-r}{\alpha}}{T_0 / T_1 + th\dfrac{R-r}{\alpha}} \right], \tag{2.4}$$

where $\sigma_0(r) = \begin{cases} \sigma_{0_1}, & 0 < R - r < 4 \text{ km}, \\ \sigma_{0_2}, & 4 < R - r < H_1 \text{ km}, \end{cases}$ $\sigma_{0_{1,2}}$ is the supposed electrical con-

ductivity of the first and second crust layers up to the depths of 4 km and $H_1 = 400$ km, respectively.

The idea of the proposed technique is finding optimal parameters of the model (2.4), which support a minimum of the misfit functional (2.5) by the gradient ascent method (Dmitriev et al., 1986):

$$\varepsilon(\mathbf{P}) = \sum_{j=1}^{L} \lg \frac{T_{j+1}}{T_j} (\lg \bar{\rho}_T^{\exp} - \lg \rho_T^{th}(\mathbf{P}))^2, \tag{2.5}$$

where $\mathbf{P}$ is the unknown parameters' vector in the model (2.4); $\rho_T^{\exp}$ are the apparent resistivities determined by means of the spherical harmonic analysis of the geomagnetic field variations for the periods from hours up to 11 years collected from all available data sources; ρ_T^{th} are theoretical values of the apparent resistivites for the model (2.4); L is the number of the periods T_j, $\lg(T_{j+1} / T_j)$ are the weighting coefficients, which take into account the nonuniform distribution of the empirical data by the periods.

The study conducted in the cited above article showed that the functional ε was quite large. The main source of this inconsistency was the significant difference between the behavior of the model and theoretical curves in the interval of long periods. In order to fix this problem a special study was carried out.

2.4.2   An Updated Formula for Deep Electrical Conductivity

It was established in Dmitriev et al. (1986) that, in order for the discrepancy between the model and experimental curve to be insignificant, it suffices to update the formula (2.4). Evidently, this discrepancy would have been minimal had it been possible to use a more general radial dependence of conductivity, which can be cast in the following form:

$$\sigma(r) = \sigma_0(r) + \gamma(r)\exp\left[-\frac{E(r)}{2kT_1}\frac{1 - th((R-r)/\alpha)}{T_0/T_1 + th((R-r)/\alpha)}\right]. \qquad (2.6)$$

However, the data (e.g., $E(r)$) are determined highly unreliably and have a large scatter. Considering this, the following simplifications in formula (2.6) were assumed:

$$\gamma(r) = \begin{cases} \gamma_1, & H_1 < R-r < H_2, \\ \gamma_2, & H_2 < R-r < R_{core}, \end{cases} \quad E(r) = \begin{cases} E_1, & H_1 < R-r < H_2 \\ E_2, & H_2 < R-r < R_{core}. \end{cases},$$

where $H_{1,2}$ are the depths of the middles of the first and second jumps in electrical conductivity (Dmitriev et al., 1977a, 1977b), $\gamma_{1,2}$ are the conductivities of the layers confined between the depths of H_1 and H_2, H_2 and R_{core} (radius of the earth core), accordingly, and $E_{1,2}$ are the correspondent activation energies, respectively.

By introducing additional parameters $H_{1,2}$ and $\gamma_{1,2}$, the probable sharp changes in the electrical conductivity in the layer C of the earth's mantle (the so-called conductivity jumps) were allowed for, which were supposed by many authors based on the physical and mineralogical arguments (Banks, 1969, 1972; Schmucker, 1970; Hobbs, 1983). Using the technique for determining the optimal parameters (Dmitriev et al., 1986), the following electrical conductivity model was obtained:

$$\sigma(r) = \begin{cases} \sigma_{0_1} = 1\,S/m, 0 < R-r < 4\ km \\ \sigma_{0_2} = 10^{-3}\,S/m, 4 < R-r < H_1 = 470\ km, \\ \gamma_1 \exp\left[-E_1/2kT_1\dfrac{1 - th\dfrac{R-r}{\alpha}}{T_0/T_1 + th\dfrac{R-r}{\alpha}}\right], H_1 < R-r < H_2, \\ H_2 = 1650\ km, E_1 = 2.4\ eV, \\ \gamma_2 \exp\left[-E_2/2kT_1\dfrac{1 - th\dfrac{R-r}{\alpha}}{T_0/T_1 + th\dfrac{R-r}{\alpha}}\right], H_2 < R-r < R_{core}, \\ T_1 = 3400\ K, \\ \alpha = 870, \gamma_1 = 8.0\,S/m, E_2 = 5.1\ eV, \gamma_2 = 47\,S/m, \end{cases} \qquad (2.7)$$

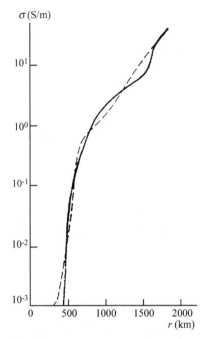

FIGURE 2.7 The deep geoelectrical profile: solid line indicates the electrical conductivity distribution corresponding to model (2.7); the dashed line corresponds to model II from (Dmitriev et al., 1986).

The electrical conductivity model itself and the ρ_T curve corresponding to the model (2.7) are shown in Figures 2.7 and 2.8, respectively.

Dmitriev et al. (1977b) have carried out the resolution study for the parameters of the model (2.7). It indicates that a 10% variation in the parameters α, H_1, T_1, $E_{1,2}$ causes ε (**P**) (2.5) to change by 100%, while other parameters, if changed by 30–50%, provide a twofold increase of the functional.

2.4.3 The Temperature Model of the Mantle

Dmitriev et al. (1988) have built the global temperature profile from the electrical conductivity model (2.7) (Figure 2.9, curve 1).

Note that the obtained distribution $T(r)$ lies between the adiabatic curve 2 and melting curve 3, which testifies in favor of the convective heat transfer in the mantle. Thus, if we assume that the distribution of electrical conductivity in the class of parametric models is described by function (2.6), then, in the case of electronic conduction mechanism, it is possible to determine the temperature profile $T(r)$ in the transient layer C and lower mantle from the global MVS data.

Traditionally, the calculations of distribution $T(r)$ from the electrical conductivity data were carried out by formula $T = (-E/2k)/\ln(\sigma/\sigma_0)$. In these calculations, in addition to the electrical conductivity and activation energy, it is also required to know the value of σ_0, which is determined by the composition

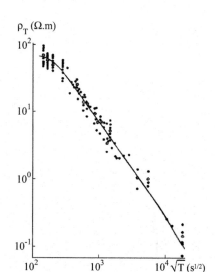

FIGURE 2.8   The global data of apparent resistivity ρ_T (dots) and its distribution corresponding to the conductivity model (2.7) (solid line) (after Dmitriev et al., 1986).

of the mantle. The estimates of this parameter from the laboratory experiments are barely reliable. The suggested method for reconstructing function $T(r)$ avoids this limitation, since in this approach the vector of the unknown parameters **P** is directly found from the experimental data of the global MVS. The data of global MVS are not just suitable for reconstructing the conductivity profile $\sigma(r)$. These data also make it possible to estimate the depth variations of temperature in the transition layer C and in the upper part of the lower mantle and to estimate the activation energy in these areas.

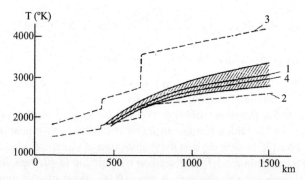

FIGURE 2.9   The depth distribution of the temperature (Dmitriev et al., 1988): 1 – the temperature distribution revealed from the electrical conductivity model (2.7); 2 – the adiabatic distribution of (Zharkov, 1983); 3 – the melting curve according to (Zharkov, 1983); 4 – the temperature distribution from (Zharkov, 1958); the hatched area shows the probable temperature range according to the global sounding data with the account of the resolution of the optimal parameters.

2.5 CONCLUSIONS

Generally, due to the complex heterogeneous structure of the earth and insufficient data on the interdependences in the multidimensional space of the parameters, the indirect geothermometers discussed above only provide the coarse temperature estimates at characteristic depths in the earth. Close relations between the electrical conductivity and temperature values (tabulated or expressed as empirical formulas valid under specific conditions or attributed to some geographical area) enable, in principle, to reconstruct the temperature profiles from the laboratory studies of rock samples collected in the study area or, at the lack of this data, from the electromagnetic sounding results expressed as the electrical conductivity (or resistivity) models.

In turn, in the latter case there are several options depending on the type and volume of the prior information. For converting the resistivity values derived from the electrical sounding into the temperature ones without study of the rock samples *in situ*, it is required to have at least a petrological information or guess about the lithology. Alternatively, the temperature profile could be reconstructed from the electrical conductivity profile if one has a guess about the conductance mechanism and appropriate activation energy. In this case it is possible to determine the parameters of the guessed electrical conductivity–temperature analytical dependence from the electromagnetic sounding data by solving of an appropriate optimization task.

It is important mentioning that the techniques considered above are inefficient for reconstructing temperature models (especially, 2-D or 3-D) at crustal depths where it is required to estimate the temperature with higher accuracy and resolution than at large depths. In the next chapter we will introduce an alternative method for building the models of crustal temperature, based on the neural network approach. It does not require prior information on the lithology, properties of rock samples, guess about the possible conductivity–temperature dependence, conductance mechanisms, and information about the appropriate activation energies.

REFERENCES

Adam, A. (Ed.), 1976. Geoelectric and geothermal studies (East-Central Europe, Soviet Asia). Akademia Kiado, Budapest, Hungary.

Alexidze, M.A., Gugunava, G.E., Kiria, J.K., Chelidze, T.L., 1993. A three-dimensional geothermal stationary model of thermal and thermoelastic fields of the Caucasus. Tectonophysics 227, 191–202.

Arnorsson, S., Gunnlaugsson, E., 1985. New gas geothermometers for geothermal exploration-calibration and application. Geochim. Cosmochim. Acta 49 (6), 1307–1325.

Arps, J.J., 1953. The effect of temperature on the density and electrical resistivity of sodium chloride solutions. Trans. Am. Inst. Mining Metallurg. Petrol. Eng. 198, 327–330.

Artemieva, I.M., 2006. Global $1° \times 1°$ thermal model TC1 for the continental lithosphere: implications for lithosphere secular evolution. Tectonophysics 416 (1–4), 245–277.

Banks, R.J., 1969. Geomagnetic variations and the electrical conductivity of the upper mantle. Geophys. J. Roy. Asron. Soc. 17 (1–4), 457–487.

Banks, R.J., 1972. The overall conductivity distribution of the Earth. J. Geomag. Geoelectr. 24 (3), 337–351.

Berdichevsky, M.N., Dmitriev, V.I., Golubtsova, N.S., Mershchikova, N.A., Pushkarev, P. Yu., 2007. In: Spichak, V.V. (Ed.), Electromagnetic sounding of the Earth's interior. Elsevier, Amsterdam, Boston, Heidelberg, pp. 27–54.

Björnsson, A., 2008. Temperature of the Icelandic crust: inferred from electrical conductivity, temperature surface gradient, and maximum depth of earthquakes. Tectonophysics 447, 136–141.

Bondarenko, A.T., 1968. Generalization of data on the conductance of igneous rocks at high temperatures in connection with the structure of the crust and upper mantle. Dokl. Akad. Nauk. SSSR 178, 20–22.

Caldwell, G., Pearson, C., Zayadi, H., 1986. Resistivity of rocks in geothermal systems – a laboratory study (Expanded Abstr). 8th NZ Geothermal Workshop, pp. 227–231.

Cermak, V., Lastovickova, M., 1987. Temperature profiles in the Earth of importance to deep electrical conductivity models. PAGEOPH 125, 255–284.

Clauser, C., 2009. Heat transport processes in the Earth's crust. Surv. Geophys. 30, 163–191.

Constable, S., Shankland, T.J., Duba, A., 1992. The electrical conductivity of an isotropic olivine mantle. J. Geophys. Res. 97, 3397–3404.

Coster, H.P., 1948. The electrical conductivity of rocks at high temperature. Monthly Notices of the Royal Astronomical Society. Geophys. Suppl. 5, 193–199.

Dakhov, V.N., 1962. Geophysical well logging. Quart. Colorado School Mines 57, 85–127.

Dmitriev, V.I., Rotanova, N.M., Zakharova, O.K., 1975. Building of the mathematical models of the electrical conductivity and temperature distributions in the Earth's interior. In: Pushkov, N.V. (Ed.), Analysis of the space-time structure of the geomagnetic field. Nauka Publ, Moscow, pp. 111–129, (in Russian).

Dmitriev, V.I., Rotanova, N.M., Zakharova, O.K., 1988. Estimation of the temperature distribution in the transient layer and the lower mantle based on the global magnetovariational data. Izvestiya Phys. Solid Earth 2, 3–8.

Dmitriev, V.I., Rotanova, N.M., Zakharova, O.K., Balikina, O.N., 1977a. Geoelectrical and geothermal interpretation of the global magnetovariational sounding results. Geomagnet Aeronom 2, 315–321.

Dmitriev, V.I., Rotanova, N.M., Zakharova, O.K., Balikina, O.N., 1986. The deep electrical conductivity model revealed from generalized results of the magnetovariational sounding. Geomagnet. Aeronom. 26, 299–306.

Dmitriev, V.I., Rotanova, N.M., Balikina, O.N., Zakharova, O.K., 1977b. On the resolving power of the global magnetovariational sounding profiles. Geomagnet. Aeronom. 6, 1092–1097.

Duba, A., Heard, H.C., Schock, R.N., 1974. Electrical conductivity of olivine at high pressure and under controlled oxygen fugacity. J. Geophys. Res. 79, 1667–1673.

Eppelbaum, L., Kutasov, I., Pilchin, A., 2014. Applied geothermics. Lecture notes in Earth system sciences. Springer-Verlag, Berlin, Heidelberg.

Flóvenz, O.G., Georgsson, L.S., Árnason, V., 1985. Resistivity structure of the upper crust in Iceland. J. Geophys. Res. 90, 10136–10150.

Foulger, G.R., 1995. The Hengill geothermal area, Iceland: variation of temperature gradients deduced from the maximum depth of seismogenesis. J. Volcanol. Geotherm. Res. 65, 119–133.

Frischknecht, F.C., 1967. Fields about an oscillating magnetic dipole over a two-layered earth and application of ground and airborne surveys. Colorado School Mines Quart. 62, 1–326.

Gaillard, F., 2004. Laboratory measurements of electrical conductivity of hydrous and dry silicic melts under pressure. Earth Planet. Sci. Lett. 218 (1/2), 215–228.

Haak, V., 1980. Relations between electrical conductivity and petrological parameters of the crust and upper mantle. Geophys. Surv. 4, 57–69.

Harinarayana, T., Azeez, A., Murthy, K.K., Veeraswamy, D.N., Rao, K.E., Manoj, V., Naganjaneyulu, V., 2006. Exploration of geothermal structure in Puga geothermal field, Ladakh Himalayas, India, by magnetotelluric studies. J. Appl. Geophys. 58 (4), 280–295.

Harvey, C.C., Browne, P.R.L., 1991. Mixed-layer clay geothermometry in the Wairakei geothermal field39Clay and Clay Minerals, New Zealand, pp. 614–621.

Hermance, J.F., Grillot, L.R., 1970. Correlation of magnetotelluric, seismic and temperature data from Southwest Iceland. J. Geophys. Res. 75 (32), 6582–6591.

Hermance, J.F., Grillot, L.R., 1974. Constraints on temperatures beneath Iceland from magnetotelluric data. Phys. Earth Planet. Inter. 8, 1–12.

Hobbs, B.A., 1983. Inversion of broad frequency band geomagnetic response data. J. Geomagn. Geoelectr. 3B (11-12), 723–732.

Hoffmann-Rothe, A., Ritter, O., Haak, V., 2001. Magnetotelluric and geomagnetic modelling reveals zones of very high electrical conductivity in the upper crust of Central Java. Phys. Earth Planet. Inter. 124 (3-4), 131–151.

Jones, A.G., 1999. Imaging the continental upper mantle using electromagnetic methods. Lithos 48, 57–80.

Jones, A.G., Lezaeta, P., Ferguson, I.J., Chave, A.D., Evans, R.L., Garcia, X., Spratt, J., 2003. The electrical structure of the Slave craton. Lithos 71, 505–527.

Karato, S., 1990. The role of hydrogen in the electrical conductivity of the upper mantle. Nature 357, 272–273.

Kharaka, Y.K., Mariner, R.H., 1989. Chemical geothermometers and their application to formation waters from sedimentary basins. In: Naeser, N.D., McCulloch, T. (Eds.), Thermal history of sedimentary basins: S. C. P. M. special issue. Springer Verlag, pp. 99–117.

Korja, T., 1993. Electrical conductivity distribution of the lithosphere in the central Fennoscandian Shield. Precambrian Res. 64, 85–108.

Korja, T., Engels, M., Zhamaletdinov, A.A., 2002. Crustal conductivity in Fennoscandia – a compilation of a database on crustal conductance in the Fennoscandian Shield. Earth Planets Space 54 (5), 535–558.

Kristinsdottir, L.H., Flóvenz, O.G., Arnason, K., Bruhn, D., Milsch, H., Spangenberg, E., Kulenkampff, J., 2010. Electrical conductivity and P-wave velocity in rock samples from high-temperature Icelandic geothermal fields. Geothermics 39, 94–105.

Kurtz, R.D., Craven, J.A., Niblett, E.R., Stevens, R.A., 1993. The conductivity of the crust and mantle beneath the Kapuskasing Uplift: electrical anisotropy in the upper mantle. Geophys. J. Int. 113, 483–498.

Lebedev, E.B., Khitarov, N.I., 1964. Dependence on the beginning of melting of granite and the electrical conductivity of its melt on high water vapor pressure. Geochem. Int. 1, 195–201, (in Russian).

Ledo, J., Jones, A.G., 2005. Upper mantle temperature determined from combining mineral composition, electrical conductivity laboratory studies and magnetotelluric field observations: application to the intermontane belt, northern Canadian Cordillera. Earth Planet. Sci. Lett. 236, 258–268.

Limberger, J., van Wees, J.D., 2013. 3D subsurface temperature model of Europe for geothermal exploration (Expanded Abstr). EAGE Conference, Amsterdam, The Netherlands.

Llera, P.J., Sato, M., Nakatsuka, K., Yokoyama, H., 1990. Temperature dependence of the electrical resistivity of water-saturated rocks. Geophysics 55, 576–585.

Lyubimova, E.A., 1968. Thermics of the Earth and Moon. Nauka Publ, Moscow, 279 pp. (in Russian).

Mareschal, M., Keller, R.L., Kurtz, R.D., Ludden, J.N., Ji, S., Bailey, R.C., 1995. Archaean cratonic roots, mantle shear zones and deep electrical anisotropy. Nature 375, 134–137.

Neal, S.L., Mackie, R.L., Larsen, J.C., Schultz, A., 2000. Variations in the electrical conductivity of the upper mantle beneath North America and the Pacific Ocean. J. Geophys. Res. 105, 8229–8242.

Olhoeft, G.R., Ucok, H., 1977. Electrical resistivity of water saturated basalt. EOS 50, 1235.

Ollinger, D., Baujard, C., Kohl, T., Moeck, I., 2010. 3-D temperature inversion derived from deep borehole data in the Northeastern German Basin. Geothermics 39, 46–58.

Parkhomenko, E.I., 1967. Electrical properties of rocks. Plenum Press, New York.

Podgornykh, L.V., Gramberg, I.S., Khutorskoi, M.D., Leonov, Yu.G., 2001. Three-dimensional geothermic model of the Karskii shelf and forecasting of the oil-and-gas presence. Dokl. Acad. Nauk 390 2, 228–232.

Polyak, B.G., Tolstikhin, I.N., 1985. Isotopic composition of the Earth's helium and the motive forces of tectogenesis. Chem. Geol. 52, 9–33.

Pommier, A., Gaillard, F., Malki, M., Pichavant, M., 2010. Methodological re-evaluation of the electrical conductivity of silicate melts. Amer. Mineral. 95, 284–291.

Pribnow, D., Hamza, V., 2000. Enhanced geothermal systems: new perspectives for large scale exploitation of geothermal energy resources in South America (Expanded Abstr). XXXI International Geological Congress. Rio-de-Janeiro, Brasil.

Rai, C.S., Manghnani, M.H., 1978a. Electrical conductivity of ultramafic rocks to 1820°K. Phys. Earth Planet. Inter. 17, 6–13.

Rai, C.S., Manghnani, M.H., 1978b. Electrical conductivity of basalts to 1550°C. Dick, H.J.B. (Ed.), Proceedings of Chapman conference on partial melting Earth's upper mantle, 96, Bull. Oreg. Pep. Miner. Geol. Ind., p. 296.

Revil, A., Cathles, III, L.M., Losti, S., 1998. Electrical conductivity in shaly sands with geophysical applications. J. Geophys. Res. 103 (B10), 23925–23936.

Rikitake, T., 1966. Electromagnetism and the Earth's Interior. Elsevier.

Roberts, J.J., 2002. Electrical properties of microporous rock as a function of saturation and temperature. J. Appl. Phys. 91 (3), 1687–1694.

Safonov, A.S., Bubnov, V.M., Sysoev, B.K., Chemyavsky, G.A., Chinareva, O.M., Shaporev, V.A., 1976. Deep magnetotelluric surveys of the Tungus Syneclise and on the West Siberian Plate. In: Adam, A. (Ed.), Geoelectric and geothermal studies. Akademia Kiado, Budapest, Hungary, pp. 666–672.

Schmucker, V., 1970. An introduction to induction anomalies. J. Geomagn. Geoelectr. 22, 9–93.

Schultz, A., Kurtz, R.D., Chave, A.D., Jones, A.G., 1993. Conductivity discontinuities in the upper mantle beneath a stable craton. Geophys. Res. Lett. 20, 2941–2944.

Shankland, T., Ander, M., 1983. Electrical conductivity, temperatures, and fluids in the lower crust. J. Geophys. Res. 88 (B11), 9475–9484.

Shankland, T.J., Waff, H.S., 1977. Partial melting and electrical conductivity anomalies in the upper mantle. J. Geophys. Res. 82, 5409–5417.

Singh, R.P., Kant, Y., Vanyan, L., 1995. Deep electrical conductivity structure beneath the southern part of the Indo-Gangetic plains. Phys. Earth Planet. Inter. 88, 273–283.

Spichak, V.V., 2006. Estimating temperature distributions in geothermal areas using a neuronet approach. Geothermics 35, 181–197.

Spichak, V.V., Zakharova, O.K., 2012. The subsurface temperature assessment by means of an indirect electromagnetic geothermometer. Geophysics 77 (4), WB179–WB190.

Spichak, V.V., Zakharova, O.K., Goidina, A.G., 2013. A new conceptual model of the Icelandic crust in the Hengill geothermal area based on the indirect electromagnetic geothermometry. J. Volcanol. Geotherm. Res. 257, 99–112.

Stacey, F., 1977. Physics of the Earth. J. Wiley & Sons, New York.

Tesauro, M., Kaban, M.K., Cloetingh, S.A.P.L., 2008. EuCRUST-07: A new reference model for the European crust. Geophys. Res. Lett. 35 (5), 1–5.

Tozer, D.C., 1959. The electrical properties of the Earth's interiors. Phys. Chem. Earth 3, 414–436.

Tryggvason, A., Rognvaldsson, S.Th., Flóvenz, O.G., 2002. Three-dimensional imaging of P- and S-wave velocity structure and earthquake locations beneath Southwest Iceland. Geophys. J. Int. 151, 848–866.

Tyburczy, J.A., Fisler, D.K., 1995. Electrical properties of minerals and melts. In: Ahrens, T.J. (Ed.), Mineral physics and crystallography: a handbook of physical constants. American Geophysical Union, Washington, DC, pp. 185–208.

Ucok, H., Ershaghi, I., Olhoeft, G.R., 1980. Electrical resistivity of geothermal brines. J. Petrol. Tech. 32, 717–727.

Ussher, G., Harvey, C., Johnstone, R., Anderson, E., 2000. Understanding the resistivities observed in geothermal systems (Expanded Abstr). World Geothermal Congress, Kyushu-Tohoku, Japan, pp. 1915–1920.

Vanyan, L.L., Berdichevsky, M.N., Fainberg, E.B., 1977. Study of asthenosphere of East European platform by electromagnetic sounding. Phys. Earth Planet. Inter. 14 (2), P1–P2.

Vanyan, L.L., Cox, C.S., 1983. Comparison of deep conductivities beneath continents and oceans. J. Geomagn. Geoelectr. 35 (11-1), 805–809.

Volarovich, M.P., Parkhomenko, E.I., 1976. Electrical properties of rocks at high temperatures and pressures. In: Adam, A. (Ed.), Geoelectric and geothermal studies (East-Central Europe, Soviet Asia). Akademia Kiado, Budapest, Hungary, pp. 321–369.

Waff, H.S., 1974. Theoretical considerations of electrical conductivity in a partially molten mantle and implications for geothermometry. J. Geophys. Res. 79, 4003–4010.

Wannamaker, P.E., 1986. Electrical conductivity of water-undersaturated crustal melting. J. Geophys. Res. 91 (B6), 6321–6327.

Wu, X., Ferguson, I.J., Jones, A.G., 2002. Magnetotelluric response and geoelecrric structure of the Great Slave Lake shear zone. Earth Planet. Sci. Lett. 196, 35–50.

Zharkov, V.N., 1958. On the temperature and electrical conductivity of the Earth's envelop. Geomagnet. Aeronom. 4, 458–470.

Zharkov, V.N., 1983. Internal structure of the Earth and planets. Nauka, Moscow, 413 pp. (in Russian).

Zharkov, V.N., Trubitsyn, V.P., Samsonenko, L.V., 1971. Physics of the Earth and planets. Nauka, Moscow, pp. 384 (in Russian).

Chapter 3

Neural Network Approach to the Temperature Estimation

Chapter Outline

3.1 INTRODUCTION

Estimating the temperature distribution in geothermal areas from the available temperature logs is an important task in geothermal exploration and is generally done by simple averaging of the adjacent temperature logs (Pribnow and Hamza, 2000), using their spline approximation (Kiryukhin et al., 1991) or kriging interpolation (Sugrobov, 1991). All these approaches, however, have their limitations. It is known, for example, that spline or kriging methods produce artificially smooth contour maps that conceal the characteristic features of a given structure. This drawback is most apparent when reconstructing temperature distributions in complicated anisotropic media or in the presence of sharp changes in properties. In other words, it is implicitly assumed that the domain studied is geologically homogeneous.

An alternative approach to the temperature distribution estimates could be based on artificial neural networks (ANNs) also known as neuronet methods (see Haykin, 1999, and references therein). These methods were, for example, applied by Corchado and Fyfe (1999) to determine the thermal structure of ocean water masses and proved successful in solving similar problems in other scientific fields. ANNs have been also applied in different geophysical tasks (Raiche, 1991; Poulton, 2002; Spichak, 2011, and references therein).

Electromagnetic Geothermometry. http://dx.doi.org/10.1016/B978-0-12-802210-8.00003-4

The characteristic feature of ANN, which is particularly useful for estimating temperature distributions, is their capability to display the internal structure on the basis of available data, and to reproduce it in the course of an intelligent interpolation/extrapolation scheme. Furthermore, neural networks permit us to make estimates on the basis of incomplete, noisy, interrelated, and irregularly distributed data, which is often the situation in the real world. Finally, contrary to other methods, no assumptions with regard to medium homogeneity are required. In comparative studies, the ANNs have been found to produce at least as good, and often better, predictive models than standard statistical techniques.

Koike et al. (2001) used neural kriging, a modification of ANN that can be considered as form of kriging, to analyze the temperature distribution in the Hohi geothermal area of southwestern Japan. They were particularly interested in studying the accuracy of the temperature interpolation. Using an analytical 2-D reference model, these authors showed that the method gave smaller interpolation errors than ordinary kriging.

Spichak and Goidina (2005) and Spichak (2006) have studied the feasibility of applying the neuronet approach to predict temperature distributions in the interwell space using available downhole temperature logs of different types and from different locations. Below we will discuss the main results following the latter paper.

3.2 ANN WITH A TEACHER (BACKPROPAGATION TECHNIQUE)

In estimating the temperature distribution in a generic geothermal area one of the "learning-with-a-teacher methods," that is, the error backpropagation (BP) technique (Rumelhart and McClelland, 1988; Schmidhuber, 1989; Silva and Almeida, 1990) was used. In such an approach there are two stages in the inversion procedure: training of the network, and its testing, or recognition (the inversion itself). In the learning (or training) stage, the "teacher" specifies the correspondence between chosen input and output data, which is a mechanism similar to that used when training a human. The analogy with the human brain also includes the similarity of some functional elements of the biological neural system to the nonlinear system "data – parameters of the target" modeled by ANN (its elements are also called "neurons"). In both cases, the system could be considered as an n-layer network in which every neuron of a given layer is somehow connected with the neurons of other layers. A signal comes to the input layer of neurons from outside the system, but its magnitude at the neurons of the other layers depends on the signal magnitudes and connection weights of all associated neurons of the previous layer. Moreover, similar to the biological systems, the net response of an artificial neuron is described by a nonlinear function.

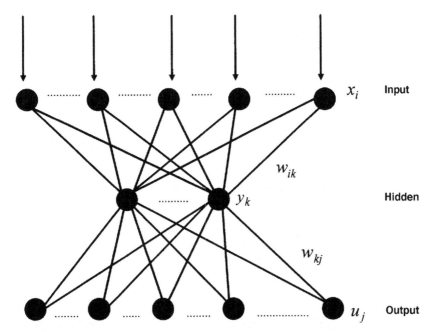

FIGURE 3.1 Example of three-layered ANN (after Spichak, 2006).

Although the BP technique has become a routine procedure, it is worth specifying here the main elements of the scheme. Figure 3.1 shows a three-layered ANN consisting of layers of input neurons (data), hidden neurons (their number, generally speaking, is arbitrary and can be adjusted in order to reflect the complexity of the system), and output neurons (unknown parameters).

Propagation of the input signal via a network occurs in the following way. The input signal x_i comes to each ith neuron of the input layer. It is equal to the correspondent element of the input vector, composed of the available temperature data. Every kth neuron of the hidden layer receives a summary input signal y_k^{inp} from all neurons of the input layer:

$$y_k^{inp} = \sum_i w_{ik} x_i,$$ (3.1)

where w_{ik} are the connection coefficients (weights) between the input and hidden layers and the summation is carried out over all input neurons. The signals y_k^{inp} are transformed by each kth neuron of the hidden layer into the output signals y_k^{out}:

$$y_k^{out} = G_k^h(y_k^{inp}),$$ (3.2)

where $G(z) = \dfrac{1}{1+e^{-z}}$ is the so-called "logistic" neuron activation function.

The signals then propagate from the hidden layer to the output layer, and for each jth neuron of the output layer we obtain:

$$u_j = G_j^u (\sum_k w_{kj} y_k^{out}),$$ (3.3)

where u_j are the output signals generated by the neurons at the output layer, w_{kj} the connection weights between the hidden and output layers, and G_j^u the activation functions for the output layer neurons. The activation functions are usually considered to be the same for each neuron of a given layer. It is worth mentioning that choosing individual activation function types for different output neurons may improve the recognition of unknown parameters (Spichak and Popova, 2000).

At the training stage the actual output signals u_j are compared with known "correct answers" u_j^t, which correspond to given input signals. A standard error

$$Er_p = \sum_j (u_{p,j} - u_{p,j}^t)^2$$ (3.4)

is calculated for each pth learning sample; here, the summation is carried out over all neurons of the output layer. In this Chapter, the term "learning sample" means a pair of "point coordinates and the corresponding temperature value." Such input–output pairs are defined by the "teacher" and comprise the ANN training sequence. The total error to be minimized is:

$$Er = \left(\frac{1}{P} \sum_p Er_p \right)^{1/2},$$ (3.5)

where the summation is performed over all P learning samples.

The connection weights w_{ik} and w_{kj} are the parameters that determine the signal propagation through the network and therefore the final error. BP is actually a gradient-descent technique, minimizing the error Er by adjusting the connection weights:

$$\Delta w_{ij}^{(n)} = -\alpha \frac{\partial Er}{\partial w_{ij}},$$ (3.6)

where $\Delta w_{ij}^{(n)}$ is the increment of the weight matrix at the nth step of the iteration process, and α is a nonnegative convergence parameter called learning rate. To accelerate the process, an inertial term proportional to the weight change at the previous step $(n-1)$ is often added to the right-hand side of Eq. (3.6):

$$\Delta w_{ij}^{(n)} = -\alpha \frac{\partial Er}{\partial w_{ij}} + \beta \Delta w_{ij}^{(n-1)},$$ (3.7)

where $\beta (0 \le \beta \le 1)$ is the inertial coefficient called the "learning momentum." The momentum can speed up training in very flat regions of the error surface and suppresses the weight oscillations in steep "valleys" or "ravines" (Schiffman et al., 1992).

Learning starts using small random weight values. The input signal comes through the network to the output. The signal from the output layer is compared with the target value and the difference is calculated. If it exceeds a predetermined small number, the signal propagates back through the network to the input layer, and so on. This procedure is repeated for the whole learning dataset and ends when a user-specified threshold value Eps (Er < Eps), known as a "teaching precision," is reached.

The testing process uses the ANN interpolation and extrapolation properties. Unlike the training procedure, which requires many transits back and forth through the network, the recognition procedure requires only one passage of the recognizable signal from the input to the output layer and utilizes the connection weights specified at the learning stage. The final result can be interpreted as an estimate of the temperature values in the specified points of the spatial mesh (in particular, at different depths in the wells).

In order to assess the quality of the ANN temperature predictions (when the true result is known in advance), the relative error (Err) averaged over all testing samples is given by:

$$\text{Err} = \frac{1}{N_{\text{test}}} \sum_{p} \frac{\left| T_{\text{obs},p} - T_{\text{ANN},p} \right|}{T_{\text{obs},p}} \times 100\%, \tag{3.8}$$

where p is the number of the temperature measurement, N_{test} is the total number of measurements in the wells, and $T_{\text{obs,p}}$ and $T_{\text{ANN,p}}$ are the observed and estimated temperature values at the pth location. It is worth mentioning in this connection that the common way of the ANN forecast errors estimating in practice (when the true answer is not known in advance) consists in dividing the initial data pool into two parts (typically 1:4) followed by using the major part for ANN training and the rest data for its testing.

The ANN architecture in all experiments carried out by Spichak (2006) was as follows: an input layer that had three neurons (coordinates of nodes): two hidden layers with 20 and 15 neurons, and an output layer that had one neuron, corresponding to the temperature to be estimated. The learning rate (α) was equal to 0.01 and the momentum (β) equal to 0.9. The neural network was taught until it reached a threshold accuracy of 1%, which, according to experience with the neural network in electromagnetic exploration (Spichak and Popova, 2000), was sufficient to achieve 5–10% accuracy in the recognition of the target parameters.

3.3 TESTING OF THE ANN

3.3.1 Analytical Temperature Model of the Geothermal Reservoir

An analytical model of the spatial temperature distribution gives the best fit for checking the ANN method. A simplified analytical model was used to compute the 3-D temperature distribution in a 1000-m-sided cubic geothermal reservoir. It was assumed that the system had a linear geothermal gradient of 0.1°C/m, and a temperature of 100°C at the top and 300°C at the bottom. At the center of the reservoir there is a permeable zone Ψ_f [$x \in (400, 600)$ m, $y \in (400, 600)$ m] that channels the upward flow of hot fluids (Sugrobov, 1991):

$$T(x, y, z) = 100 + 0.1z, (x, y) \in \Psi \setminus \Psi_f \tag{3.9a}$$

$$T(x, y, z) = 100 + 200 \exp \beta (z - 2000), (x, y) \in \Psi_f, \tag{3.9b}$$

where $\Psi = \{x \in [0, 1000], y \in [0, 1000]\}$ m, $\Psi_f = \{x \in [400, 600]$ m, $y \in [400, 600]\}$ m, $\beta = \dfrac{C_b \cdot \gamma \cdot v}{\lambda}$, C_b is a fluid heat capacity coefficient, γ the fluid density, λ the thermal conductivity coefficient of water-saturated rocks, and v the fluid flow rate. The following parameter values were used in the computations: $C_b = 4.5$ kJ/kg °C, $\gamma = 865$ kg/m^3, $\lambda = 2.1$ W/°C, $v = 10^{-7}$ m/s (A.V. Kiryukhin, personal communication, 2003).

3.3.2   Effect of the Data Volume

In order to estimate the influence of the number of samples utilized for neural network training on the accuracy of the determinations of unknown temperatures in all points of the regular mesh (x_i, $i = 1,...,11$; y_j, $j = 1,...,11$; z_k, $k = 1,...,51$) extending over the volume of the geothermal zone, the ANN was consecutively taught using sets of 10, 20, 30, 40, 50, 60, 70, 80, 90, 100, and 110 temperature profiles randomly sampled (by means of a random-number generator) from 121 "logs," corresponding to all the grid points on the ground surface ($z = 0$ m). Each set consisted of temperatures computed using Eqs. (3.9a) and (3.9b).

Figure 3.2 shows the ground surface isotherms obtained by the neural network when taught using different quantities of data. It is quite clear that, when utilizing sets with more than 60 well logs (i.e., about half of all the available data) for teaching, the isotherms delineate rather well the area of increased temperatures in the central part of the "geothermal zone."

These results are confirmed by assessing the accuracy of temperature estimates over the entire area, which indicates that it depends on the size of the dataset used for teaching the neural network (Figure 3.3). The diagram in Figure 3.3 shows the dependence of the root-mean-square error of the estimates (given in percentage of the average temperature in the area under study) on the number of well logs (NW) in the teaching set. It follows from the diagram that, for NW = 1–50, the average error can be up to 25%, but for $N > 60$ it does not exceed 10–12%, decreasing to 1% for NW = 121.

Figure 3.4 illustrates the vertical distribution of the temperatures at $y = 500$ m, obtained using routine kriging interpolation (a) and neural network reconstruction (b). It can be seen that the vertical boundaries of the hot fluid

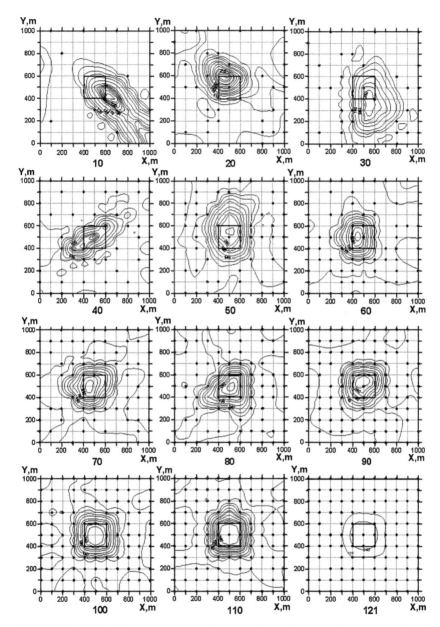

FIGURE 3.2 Maps of forecast temperature isolines on the surface depending on the number of datasets used for neuronet forecast (after Spichak, 2006). Random "well" distributions on the surface in amount of 10, 20, 30, 40, 50, 60, 70, 80, 90, 100, and 110 are plotted; "121" is a temperature contour map reconstructed by means of neural network taught by all 121 thermograms.

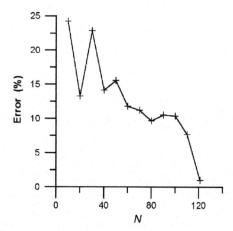

FIGURE 3.3 Plot of the dependence of a root-mean-square forecast error (in % of the average temperature value in the zone under study) on the number of thermograms (N) used for neural network teaching (after Spichak, 2006).

upflow zone [$x \in (400, 600)$ m] are more clearly defined by the results given by the neural network approach (i.e., Figure 3.4b).

Using the data from the analytical model when testing the neural network method, we can reach an assessment of its predicting capabilities (and especially how it compares with standard smoothing interpolations). In the case presented here, to achieve a 10% level of temperature prediction accuracy, it is sufficient to teach the neural network using a dataset that includes about half of the total number of well logs (i.e., 60 out of 121) for which the estimate is performed.

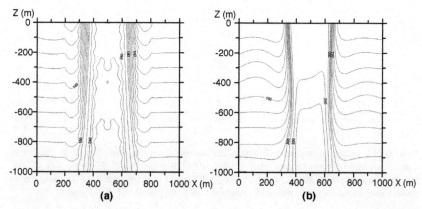

FIGURE 3.4 Vertical cross-section of the temperature distribution under the central profile ($y = 500$ m) obtained by means of kriging-interpolation (a) and neural network reconstruction (b).

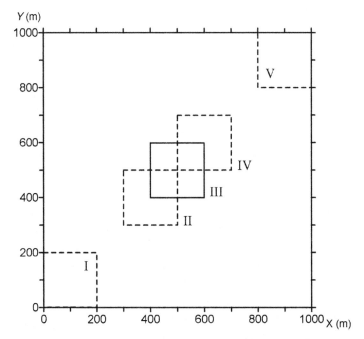

FIGURE 3.5   Surface section location used for neural network teaching and testing (after Spichak, 2006).

3.3.3   Effect of the "Geographic" Factor

Application of the neural network method to geophysical image recognition has shown that reconstruction accuracy is affected not only by the size of the teaching dataset but also by its structure (Spichak and Popova, 2000).

When the estimate is based on data measured in adjacent points, the neural network functions in the interpolation mode. Sometimes it is, however, necessary to estimate temperatures in one part of the system on the basis of well log data obtained in another. In that case the neural network functions in the extrapolation mode and the recognition results depend both on the character of the temperature distribution and the "geographic" factor (i.e., the distance between the areas from which the data used for teaching and recognition originate).

To determine the effect of the geographic factor on the accuracy of temperature estimates based on well log data, the temperatures obtained by the analytical model considered in Section 3.3.1 were used. The neural network was consecutively taught on model data from the areas confined to Squares I, II, and III shown in Figure 3.5. Estimate accuracy was assessed on the basis of the model data computed for Squares I, II, III, IV, and V, and also for the entire area. Table 3.1 presents the results obtained using the neural network approach with the same parameters as in Section 3.3.1.

TABLE 3.1 Relative forecast errors of temperature values (in %) in areas I–V (see Figure 3.5) by means of neural network taught by model data for I-III zones

Number of the zone	I	II	III	IV	V	Total area
I	1.0	28.0	34.8	27.9	20.9	31.7
II	12.0	1.0	1.8	47.3	87.9	45.2
III	67.7	46.2	1.0	46.0	68.1	63.9

After Spichak (2006).

Logically, the smallest error in the temperature predictions (1%) is achieved when the neural network is taught on data from the area for which the estimates are made (see Cells I/I, II/II, and III/III in Table 3.1). At the same time, the estimate error in Square V for a neural network taught on the model database from Square I amounts to 20%, although the temperature change with depth is similar in both areas. This is caused by the above-mentioned "geographic" factor since the measurement point coordinates are input parameters of the neural network. The more the coordinates of the measurement point differ from those of the point for which the estimate is carried out, the greater the error.

The smallest error (leaving aside the trivial example mentioned above) is achieved when estimating the temperature in the hot fluid upflow zone (Square III), using a neural network taught on data from Square II, where the temperature change with depth is of a mixed character defined by Eqs. (3.9a) and (3.9b); that is because half of Square II is in the upflow zone, the other half is outside. This can be explained by the fact that (1) one-quarter of the data in the teaching and testing samples coincide, and (2) Squares II and III are geographically close; indeed, they intersect (Figure 3.5).

It is worth mentioning at this point that, when the teaching and testing data are inverted (see Cell III/II in Table 3.1), the estimate error increases considerably to 46.2%, compared to the 1.8% for Cell II/III. This can be explained by the fact that the teaching data in this case are homogeneous (one type of temperature distribution) whereas the testing data are from two types of distribution, and because a neural network taught on the basis of one type of temperature distribution extrapolates poorly onto the other. For the same reason, the error in Case II/I is smaller than in Case I/II (12 versus 28%; Table 3.1).

Thus, on the basis of the experiments carried out on the analytical model data, one can conclude that the errors in neuronet temperature estimates using well log data are influenced by (i) the "education level" of the neural network (i.e., the proximity of the temperature distributions used for teaching to those used for recognition), and (ii) the distance between the points where estimates are made and the area providing the data used in neural network teaching.

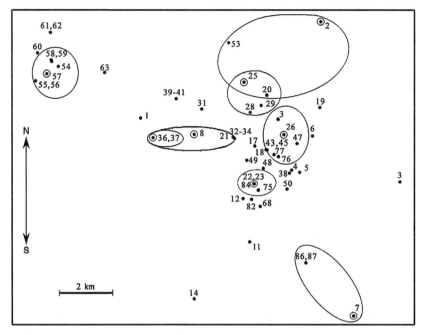

FIGURE 3.6 Map of well location in the anonymous geothermal zone (after Spichak, 2006): eight wells are marked with circles in which the temperature distribution was forecasted; areas comprising only neighbor wells are marked for each of them.

3.4 AN EXAMPLE OF THE NEURAL NETWORK BASED TEMPERATURE FORECAST IN THE GEOTHERMAL AREA

3.4.1 Data

Let us consider the application of the neuronet approach to estimating well log temperatures in an actual situation. Figure 3.6 is a map showing the location of 58 wells in one of the geothermal areas in Italy that extends 13.8 km in an east–west direction and 10.7 km in a north–south direction.

The temperatures in the wells were measured from the wellhead down to a maximum depth of 3775 m (not necessarily the true depths, since some wells were vertical). The number of measurements per well was rather small (an average of five to six per well). Figure 3.7 presents a bar chart of the distribution of the number of measurements in the wells; in almost a third of the wells only one or two temperature data points were available.

To assess the accuracy of the neuronet estimate, the 58 well logs were divided into two groups: the data used for neural network teaching (50 logs) and the data used for comparison of predicted temperatures (eight logs from Wells 2, 7, 8, 23, 25, 26, 36, and 57; the wells are indicated by the concentric circles in Figure 3.6). These eight wells were chosen because they are geographically representative of the points/areas for which temperature predictions were performed

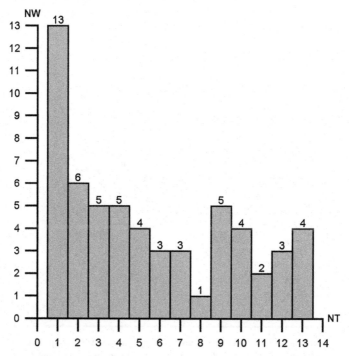

FIGURE 3.7   Histogram of the temperature records in wells: number of wells (NT) in horizontal, the number of temperature records (N) in vertical (after Spichak, 2006).

and also because their logs provide sufficient temperature data for comparisons. The choice was obviously subjective, but since the test outcome was not known beforehand, one can disregard the influence of this factor on the results.

3.4.2   Effect of the Temperature Logs' Number

As when the method was tested against the analytical model (Section 3.3), a series of experiments was carried out to determine the effect of the number of well logs used for teaching on the errors in estimated temperature for the eight wells mentioned above. The neural network was taught at first using randomly sampled logs from the teaching database.

To assess the influence of the amount of prior information on the results of the estimates, the neural network was taught consecutively, using data from 10, 20, 30, 40, and 50 randomly sampled well logs; this corresponded to 50, 140, 151, 185, and 250 measured temperatures, respectively. Finally, and to determine the effect of the geographic factor, the neural network was taught on data from wells located in the direct vicinity of each of the wells selected for testing (in Figure 3.6 these wells are grouped within ovals). The results of the experiments are presented in Table 3.2 and Figure 3.8.

TABLE 3.2 Forecast errors of eight thermograms in the geothermal zone depending on the number of thermograms (N) used for neural network teaching

N	2	7	8	23	25	26	36	57	Average
50 (250)	15.6	43.1	27.1	6.9	16.1	6.2	10.6	9.7	16.9
40 (185)	44.6	9.9	37.3	23.5	17.2	8.2	15.0	5.5	20.1
30 (151)	34.0	49.1	23.4	15.1	8.0	27.0	3.4	11.0	21.4
20 (140)	20.1	81.8	24.0	22.0	23.3	17.4	11.2	31.5	29.0
10 (50)	68.0	18.6	30.4	22.0	40.0	13.7	13.2	64.7	33.8
S	48.3	53.5	26.2	6.6	83.5	15.7	18.2	8.5	32.6
Average	38.4	25.6	28.0	16.0	31.3	14.7	11.9	21.8	

S corresponds to teaching only by thermograms from neighbor wells (see Figure 3.5) (after Spichak, 2006). The total number of temperature records in wells is given in parenthesis.

As shown in Table 3.2, when reducing the number of well logs used for neural network teaching, the estimate error averaged over all eight wells (last column) gradually increases from 16.9% (for $N = 50$) to 33.8% (for $N = 10$). The lowest errors are within the range of 3.4% (Well 36; 30 well logs in the teaching sample) to 23.4% (Well 8; 30 logs). The worst average results for this set of experiments were achieved for remote Wells 2 (38.4%; 4.3 km from the nearest well), 8 (28.0%; 1.5 km), and 7 (25.6%; 3.4 km). If these wells are excluded from consideration, then the smallest average error (last column) will decrease from 16.9 to 9.9% (not shown in Table 3.2).

It is worth noting that estimates performed by a neural network taught with the sole use of log data from adjacent wells (see the ovals in Figure 3.6, data on row S of Table 3.2, and respective plots in Figure 3.8) gave an error minimum for Well 23 of only (6.6%). Moreover, despite the fact that the temperatures for Well 36 were estimated using data from Well 37, which is only 2 m away, the error (18.2%) was higher than when using a greater number of data (see corresponding column in Table 3.2). This is because the estimate error depends mainly on the *similarity* between the temperature distributions used in teaching the network and those to be estimated, and only subordinately on the geographic factor (compare with the results obtained for the analytical model in Section 3.3.3). In other words, geographic proximity cannot compensate for a lack of similar temperature distributions in the teaching data.

The more data in the teaching sample, the higher the *probability* that the sample will include temperature distributions similar to those expected in real data. An increase in the number of well logs used in the estimates will therefore result in a decrease in the *average error over the entire ensemble* of points at which the estimates are carried out monotonically (see the last column in

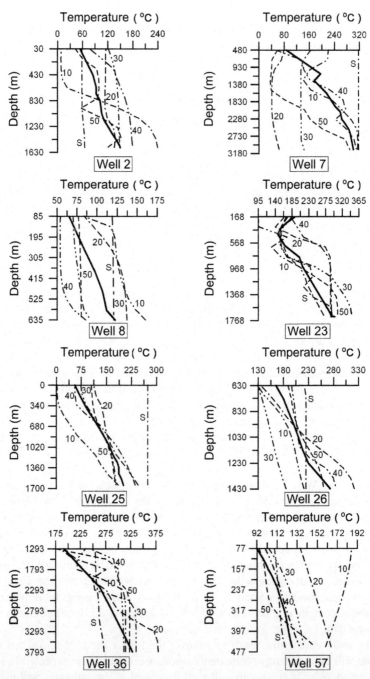

FIGURE 3.8 Forecast thermograms in eight wells marked in Figure 6.6 depending on the number of thermograms used for forecasting (10–50, S is a case of using thermograms only from neighbor wells) (after Spichak, 2006). Temperature (in °C) in horizontal, depth (in m) in vertical.

Table 3.2). At the same time, estimate *error minima and maxima for individual wells* are not directly correlated with the total amount of data used in the ANN.

Figure 3.8 presents the plots of estimated temperature versus depth for eight wells, depending on the number of well logs used for neural network teaching. The actual temperature values measured in the wells are represented by bold lines. The estimated values obtained using the neural network approach are indicated by the lines labeled 10, 20, 30, 40, and 50, which correspond to the number of well logs used for teaching; the estimates obtained only on the basis of logs from adjacent wells are labeled with the letter S – see ovals in Figure 3.6).

As follows from Figure 3.8, the best estimate curves for most wells (except the above-mentioned remote wells) are close to the measured temperature logs (bold lines). In a number of cases, only 30 logs were necessary to achieve minimal errors (Wells 25 and 36), whereas Wells 23 and 26 required 50 logs. As expected, the worst results were obtained when using only 10 and 20 logs, or only logs from adjacent wells.

One might therefore conclude that, although the average estimate error over all the test points decreases with an increase in the total amount of teaching data, the error for individual points depends to a considerable extent on the similarity of the corresponding temperature distribution to one or more in the teaching dataset. This will not necessarily be found between nearby wells, even when they are very close together. In this case, a 20% average level in the estimate error is achieved when using 30 logs, decreasing negligibly to 16.9% as their number further increases in the teaching sample (up to 50 logs). So, in the example discussed here, this amount of training data seems to characterize the system fairly well, and this kind of knowledge is important from a practical point of view.

3.4.3 Effect of the Temperature Logs' Type

It was logical to expect better temperature estimates from increases in the volume of the training database (prior information), but the influence of the structure of the database is much less obvious. It was particularly interesting to determine the accuracy of the estimates as a function of the type (conductive or convective) of temperature profile used for ANN teaching. It should be remembered that, at a conductive site, the temperature tends to increase linearly with depth, whereas at a convective site the temperature profile has an inflection point, showing, near the surface, a higher rate of temperature increase with depth.

In order to study the effect of the type of teaching sample, ANN training was performed by Spichak (2006) in three different ways, using (1) only conductive profiles, (2) only convective profiles, and (3) both types of profile (mixed sample structure). A total of 33 well logs were used in these experiments: 25 of the conductive type (four used for testing; the other 21 for teaching the neural network) and eight of the convective type (two for testing and six for teaching).

TABLE 3.3 Average forecast errors and bars (%) depending on the type of thermograms used for neural network teaching

		Type of teaching samples		
		Conductive	Convective	Mixed
Type of testing samples	Conductive	8.5 ± 3.0	14.1 ± 11.7	9.1 ± 8.4
	Convective	19.0 ± 9.3	12.3 ± 7.6	15.4 ± 7.1

After Spichak (2006).

During each experiment the test samples were randomly selected from the database. This was done five times, with the number of conductive and convective test logs remaining the same (i.e., 4 and 2, respectively).

The average values of the temperature estimates and standard deviations are summarized in Table 3.3, and indicate that the smallest estimate errors correspond to situations in which the well logs used for teaching and testing are of the same type (i.e., 8.5% in the conductive case and 12.3% in the convective case). In the latter case, the errors are negatively affected by the lack of teaching samples, compared to the situation in the former (6 versus 21), as mentioned above.

Conversely, the largest average estimate errors occur when the logs used for teaching and testing are of different types (14.1% in the convective/conductive-type case and 19.0% in the opposite case). The smaller error for the convective/conductive case is a result of teaching the ANN using nonlinear functions and testing against linear ones, which turns out to be a better approach than the other way round.

Note that teaching the ANN by using both types of log (i.e., utilizing a mixed teaching dataset) reduces the average error to 9.1 and 15.4% for the convective/conductive and conductive/convective type cases, respectively. These results indicate that use of a mixed teaching dataset practically guarantees minimal estimate errors in the case of conductive-type temperature logs, but not in the case of convective-type logs, as illustrated by the graphs in Figure 3.9.

Figure 3.9 shows the results of temperature estimates for two different types of well (i.e., Wells 18, 26, 29, and 48 are predominantly conductive, while Wells 43 and 77 show convective profiles). In all the graphs, the smallest errors are a consequence of using well logs of the same type for teaching and testing, while the largest errors correspond to cases in which different types of temperature log were used. When logs of mixed type are used, the resulting errors are, in all cases, larger than when the same type of log is used, but smaller than when different types of log are used.

Thus, when there is uncertainty with regard to which type of temperature the estimated logs correspond, the best strategy in ANN teaching is to use all available downhole temperature log data. Where temperature type is known beforehand, only well logs of similar type should be used for teaching purposes.

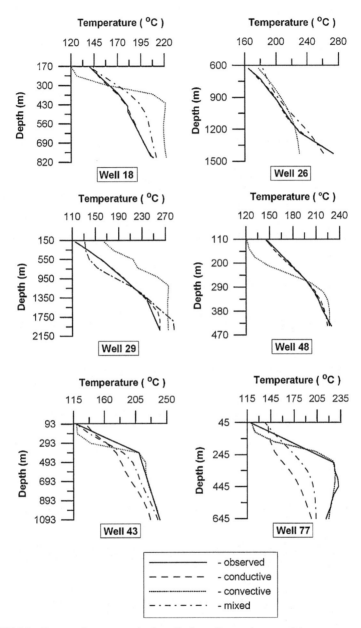

FIGURE 3.9   Forecast thermograms in six wells depending on the type of thermograms used for teaching (conductive/convective/mixed) (after Spichak, 2006). The thermograms from wells 18, 26, 29, and 48 are of conductive type, while the thermograms from wells 43 and 77 are of convective type. Temperature (in °C) in horizontal, depth (in m) in vertical.

3.5   CONCLUSIONS

It has been demonstrated that neuronet downhole temperature estimates can be made using neural networks taught by available well log data. The method was tested using 3-D analytical temperature model of the geothermal reservoir. It was shown that, in neuronet temperature estimates based on well log data, the errors are influenced by (1) the "education level" of the ANN, and (2) the distance between the estimate points, and the area for which information was available for neural network teaching.

These conclusions are confirmed by estimating well log temperatures using available log data from a geothermal area. This approach revealed that the average estimate error for all the cases studied decreases to 16.9% when the total number of data used for neural network teaching was increased (if the remote wells are excluded, this average decreases to 9.9%). Individual estimate errors depend to a considerable degree on the existence of similar temperature profiles in the teaching dataset. It should be mentioned in this connection that a further increase in the amount of training data (up to a volume corresponding to the system under study) could result in a further decrease in average estimate errors.

To determine the effect of the structure of the teaching data, a study was made of the relationship between estimate accuracy and the type of temperature profiles used in ANN teaching (conductive or convective). The best results were achieved, as expected, by using similar types of temperature well log for teaching and testing. It was less evident that a mixed teaching dataset has practically no effect on the mean estimate error when inferring conductive-type temperature logs, whereas the same dataset leads to a conspicuous increase in the mean error when estimating wells logs of the convective type. Accordingly, the optimal strategy in ANN teaching is to obtain information beforehand on the type of temperature well logs that are to be determined. It is better to use logs of a similar type as the one being estimated, and, in the absence of such knowledge, to use all the data available.

The accuracy of the estimates can be improved by as much as 5–10% if neural network teaching is aimed at achieving the lowest possible error (i.e., less than 1%, as in the experiments discussed here); this will entail a considerable increase in computation time, and may be justified when the temperature distribution has an important impact on the exploration effort in a geothermal area, such as in the location of drilling sites.

Finally, it should be noted that the ANN method proposed here could also be useful when estimating temperatures for regions where no prior geological and geophysical information is available, and without resorting to heat flow models. However, wherever these data are available, they could be incorporated in the ANN forecast at the neural network teaching stage.

In the next chapter we will demonstrate how the neural network technology could be joined with the electromagnetic sounding data so that to provide better temperature estimates.

REFERENCES

Corchado, J.M., Fyfe, C., 1999. Unsupervised neural method for temperature forecasting. Artif. Intell. Eng. 13, 351–357.

Haykin, S., 1999. Neural networks: a comprehensive foundation, second ed. Prentice Hall.

Kiryukhin, A.V., Fage, D.M., Blukke, P.P., Demchenko, A.A., Perveev, S.L., Gusev, D.N., 1991. Reconstruction of 3D temperature fields in geothermal reservoirs using spline-approximation with Green function. Volcanol. Seismol. 3, 37–48.

Koike, K., Matsuda, S., Gu, B., 2001. Evaluation of interpolation accuracy of neural kriging with application to temperature distribution analysis. Math. Geol. 33 (4), 421–448.

Poulton, M., 2002. Neural networks as an intelligence application tool: a review of applications. Geophysics 67, 979–993.

Pribnow, D., Hamza, V., 2000. Enhanced geothermal systems: new perspectives for large scale exploitation of geothermal energy resources in South America (Expanded Abstr). XXXI International Geological Congresspp. 361–362.

Raiche, A.A., 1991. Pattern recognition approach to geophysical inversion using neural networks. Geophys. J. Int. 105, 629–648.

Rumelhart, D., McClelland, J., PDP Research Group, 1988. Parallel Distributed Processing 1. MIT Press, Cambridge, 576 pp.

Schiffman, W., Joost, M., Werner, R., 1992. Optimization of the backpropagation algorithm for training multilayer perceptrons. Technical report. Institute of Physics, University of Koblenz, 36 pp.

Schmidhuber, J., 1989. Accelerated learning in back-propagation nets. Connectionism in Perspective. Elsevier Science, pp. 439–445.

Silva, F.M., Almeida, L.B., 1990. Speeding up backpropagation. In: Eckmiller, R. (Ed.), Advanced neural computers. Elsevier Science, pp. 151–158.

Spichak, V., 2006. Estimating temperature distributions in geothermal areas using a neuronet approach. Geothermics 35, 181–197.

Spichak, V., 2011. Application of ANN based techniques in EM induction studies. In: Petrovský, E., Herrero-Bervera, E., Harinarayana, T., Ivers, D. (Eds.), The Earth's magnetic interior: IAGA special Sopron book series. Springer, pp. 19–30, vol. 1.

Spichak, V.V., Goidina, A.G., 2005. Temperature prediction in geothermal zones from borehole measurements using neural networks. Izvestiya Phys. Solid Earth 41 (10), 79–88.

Spichak, V.V., Popova, I.V., 2000. Artificial neural network inversion of MT – data in terms of 3D earth macro – parameters. Geophys. J. Int. 42, 15–26.

Sugrobov, V.M. (Ed.), 1991. High-temperature hydrothermal reservoirs. Nauka, Moscow, 158 pp. (in Russian).

Chapter 4

Indirect Electromagnetic Geothermometer

Chapter Outline

4.1 GENERAL SCHEME OF THE ELECTROMAGNETIC GEOTHERMOMETER

Complex inhomogeneous structure of the earth and lacking information about appropriate dependencies in the multiparametric space allow only very coarse temperature estimations in the framework of the approaches discussed in Chapter 2. Meanwhile, due to interconnection between the temperature and electrical resistivity/conductivity (see Section 1.3.1) using of the latter data revealed from the electromagnetic sounding could reduce the temperature estimation errors since in this case the initial database is increased due to adding of a new data related somehow to the temperature. Accordingly, the electrical resisivity/conductivity could be used as a proxy parameter effective for bridging the gaps in the temperature data both in the interwell space and beneath the boreholes. In other words, the electromagnetic sounding results could be used for indirect temperature estimations in the earth.

The most effective mathematical tools capable of dealing with different types of data without prior knowledge regarding the thermal conductivity or heat flow data as well as guessing the electrical conductance mechanisms in the

Electromagnetic Geothermometry. http://dx.doi.org/10.1016/B978-0-12-802210-8.00004-6

earth's crust could be based on the supervised artificial neural network apparatus (Haykin, 1999). The latter is known as a powerful approximation tool, which does not require preliminary knowledge and could extract necessary information from the input–output data pairs used for its training (see Spichak (2011) and references therein for a review on artificial neural network (ANN) application in the EM induction studies). As it was shown in Chapter 3, ANN provides more accurate temperature estimates in the interwell space than, say, ordinary kriging.

Technique of the indirect temperature estimations in the earth interior from the electrical conductivity/resistivity (revealed from EM sounding data) using a supervised neural network is called an "Indirect Electromagnetic Geother-mometer." It was first proposed by Spichak et al. (2007) and further developed in Spichak and Zakharova (2009a, 2009b, 2010, 2011; Spichak et al. (2011).

The application of the indirect EM geothermometer consists of the follow-ing steps:

- EM sounding of the study area;
- EM data inversion resulting in the electrical conductivity/resistivity model up to the required depth;
- geothermometer calibration using available temperature logs and conductivity/ resistivity data;
- ANN temperature reconstruction in the predetermined area from its electri-cal conductivity/resistivity model.

Since the EM sounding techniques as well as the data inversion methods are well known (see, for instance, Spichak, 2007), we will restrict ourselves by only considering of the latter two issues: geothermometer calibration and the temperature reconstruction.

At the calibration stage training of the supervised artificial neural network in the correspondence between the data of the electrical conductivity/resistivity profiles and the temperature logs from the adjacent wells is carried out. (The mathematical process is described by formulas 3.1–3.8 from Section 3.2.) In this case the ANN input consists from the electrical conductivity/resistivity values and the appropriate space coordinates, where they are determined, while the output are the temperature values determined in the same locations. In practice this could be realized by advanced estimating of the conductivity/ resistivity values (in particular, by ANN) in the locations where the tempera-ture is recorded.

After the geothermometer is calibrated the corresponding ANN is ready to be used for the temperature estimation in the area of interest. In this case the ANN input consists from the resistivity/conductivity values taken from the model, while the output consists from the temperature estimates in the same spatial locations.

In the next sections we will discuss the methodological issues of the temper-ature interpolation in the interwell space and extrapolation in depth by means of the indirect EM geothermometer.

4.2   EM TEMPERATURE INTERPOLATION IN THE INTERWELL SPACE

Parameter estimation in the space between the drilled boreholes is usually carried out by linear interpolation or geostatistical tools based on the spatial statistical analysis of the approximated function, "kriging" being the most often used procedure (Chiles and Delfiner, 1999). As it was shown in Section 3.3 using of the ANN technique enables to reduce the interpolation errors while application of the indirect EM geothermometer further reduces these errors (Spichak et al., 2011). Below we will consider the methodological issues of the calibration of the EM geothermometer following this paper.

4.2.1   Effect of the Data Volume

As it is known (Haykin, 1999), the quality of the ANN reconstruction depends on the training data volume and representativeness. The effect of the data volume used for the EM geothermometer calibration was studied by Spichak et al. (2007, 2011). Magnetotellric and temperature data used for the thermometer calibration were collected in the northern Tien Shan study area (Figure 4.1). MT data were measured in the frequency range from 5×10^{-4} to 300 Hz in the sites located in the vicinity of eight boreholes (T4, T5, T7, T8, T9, T10, T11, T12), where the temperature was measured up to the depths exceeding 1 km (Duchkov et al., 2001).

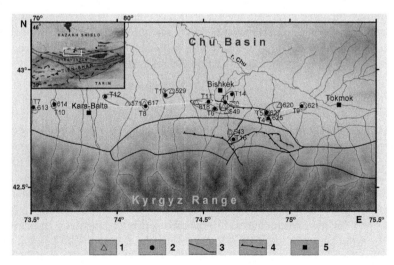

FIGURE 4.1   Location of MT sites and boreholes for which temperature data are available in the Tien Shan area (Spichak et al., 2011): 1 – MT sites; 2 – boreholes; 3 – deep faults; 4 – thrust faults; 5 – large cities. Shadowing indicates the topography of the area. MT-site-borehole pairs used for the indirect geothermometer calibration are encircled. The white line connects the MT sites used for EM thermometer calibration (see Section 5.5.2 of the Chapter 5).

In order to reveal 1-D electrical conductivity profiles from magnetotelluric data, the Bostick inversion of the impedance "determinant" was used, which is known to be robust with respect to multidimensional disturbances (e.g., Park and Livelybrooks, 1989). Figure 4.2 shows 15 pairs of the apparent electrical conductivity and temperature profiles from the nearest boreholes.

During the feasibility studies the total database each time was divided into two parts used for training and testing of the ANN. To estimate the effect of the training sample size, neuronets were successively trained with data from 2, 4, 6, 8, 10, and 12 pairs of adjacent temperature and electrical conductivity profiles (here and after "T–MT") that were randomly selected from the total dataset. Accordingly, the taught ANNs were tested on the rest pairs of profiles. During the testing the temperatures were forecasted in the locations of the temperature records in the wells not used for training and compared with real temperature logs. The appropriate misfits were determined by the formula (3.8) from Chapter 3.

For comparison, other neuronets were trained using temperature logs alone. The relative *rms* errors of temperature predictions in the borehole locations are plotted in Figure 4.3 for the cases of temperature data alone (line with triangles) and electromagnetic data paired with temperature logs (dotted line).

Comparison of the graphs shows that, if both temperature and electromagnetic data are used for temperature prediction, an increase in the training sample size decreases the relative error more rapidly than if temperature logs are used alone. Moreover, the prediction error reaches a minimum value for a sample consisting only of six T–MT pairs, whereas the estimation from temperature logs attains the same level with the use of data of eight to 10 boreholes. The important implication of this is that if borehole measurements of temperature are limited, the temperature prediction error can be substantially reduced (by nearly two times) by using both temperature and EM sounding data.

4.2.2   Effect of the Neuronet Training Strategy

To examine the effect of the neuronet training strategy on the error of temperature prediction from electromagnetic data, two strategies were used. In the first case, neuronets were trained with five samples of 12 randomly selected pairs of T–MT profiles, and the temperatures in three boreholes whose data were not used for training were then predicted from the electrical conductivity data of the nearest MT sites. In doing so, the temperatures in boreholes T5 and T6 (see Figure 4.1 for their locations) were predicted separately for the conductivity profiles from sites 627 and 618 (T5) and from sites 620 and 549 (T6). On the other hand, the electromagnetic data at MT sites 618 and 550 were analyzed together with temperature profiles measured not only in boreholes T6 and T1, but also in T11 and T14, respectively.

In the framework of the second strategy, the neuronet was trained with all available MT data, after which it was used for predicting the electrical

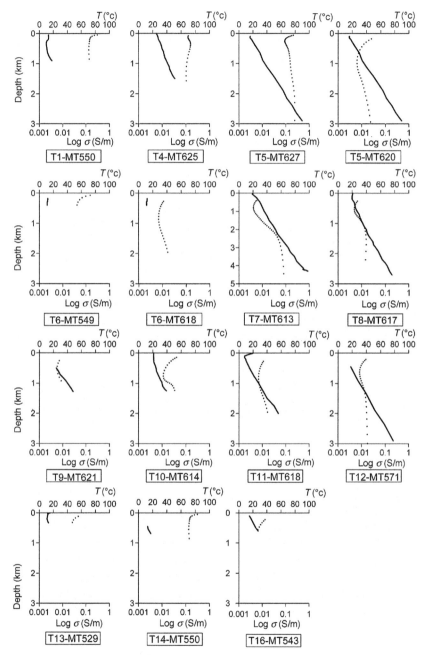

FIGURE 4.2   Temperature well logs (solid lines) and electrical conductivity profiles beneath adjacent MT sites (dotted lines) in the Tien Shan area (after Zakharova et al., 2007).

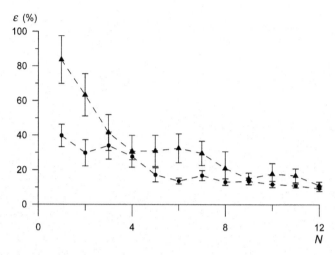

FIGURE 4.3 Average relative error ε (in %) of the temperature estimation in the Tien Shan area based on the data on electrical conductivity (dots) and temperature logs (triangles) as a function of the number of pairs (N) of temperature and electrical conductivity profiles (or temperature logs only) used for the neuronet training (after Spichak et al., 2011).

conductivity in the locations of the borehole temperature records. Finally, the neuronet trained on the basis of the correspondence between electrical conductivity and temperature in 14 pairs of T–MT profiles was used to predict the temperature in a borehole whose data were not used for training. In order to compare the results of temperature prediction based on electromagnetic and temperature data with results obtained by means of neuronets trained with temperature data alone, neuronets were trained only with the same temperature logs, and the temperatures in the same boreholes were predicted.

The predicted results are presented in Figure 4.4 and left three columns of Table 4.1. The uncertainties of the temperature prediction by the first and second techniques (the use of electrical conductivity data from the nearest MT site and the "blind" use of all available MT data) are given in columns 1 and 2 of Table 4.1, respectively. The uncertainties of the ANN prediction from the third method (temperature logs alone) are given in column 3.

The average relative error of the temperature prediction evaluated by the first technique was 11.9%, which is an unexpectedly good result for this region, which is characterized by a complex geological structure and a large scattering of temperature distributions (Zakharova et al., 2007). The average relative errors of prediction by the second and third methods were 29.9 and 29.5%, respectively. Although the prediction errors of the second and third techniques were in three cases smaller than those of the first one, the results predicted by the first technique were better in 80% of the cases.

In other words, a reasonable choice of EM sites as close as possible to the points *at which the temperature is to be predicted* yields the best results (see Section 4.2.3 for the estimates of the appropriate spacing affect). However, the

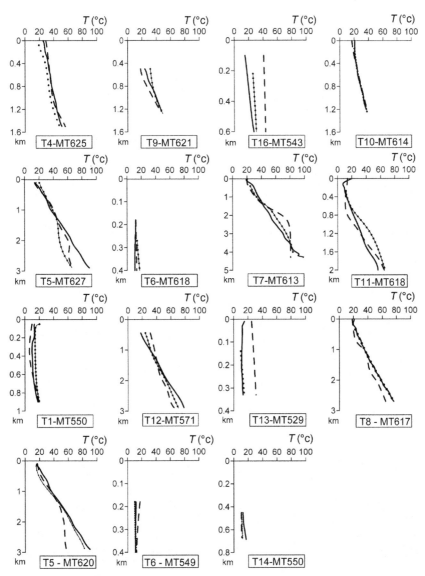

FIGURE 4.4 Measured and estimated temperature profiles in wells located in the Tien Shan area (Spichak et al., 2011). Solid line – measured temperature, dashed line – temperature model based on the temperature data; line with dots – temperature model based on MT data.

distance between the temperature prediction point, on the one hand, and pairs of EM sites and the corresponding boreholes whose data were used for calibration, on the other hand, is not a decisive factor. This is evident from the comparison of prediction results obtained by horizontal extrapolation to boreholes T7, T9, and T16, which were located on the periphery of the region studied. Table 4.1

TABLE 4.1 Errors of the borehole temperature estimation as dependent on the calibration strategy, borehole-MT site spacing, local geology, and hydrological conditions in the Tien Shan area

Boreholes and related MT sites	Relative errors of the temperature estimation (%)			Borehole-MT site spacing (km)	Factors influencing the temperature estimation errors			
	1	2	3		Spacing (> 2 km)	Faulting	2-D/3-D heterogeneity	Meteoric water flows
T1-MT550	24.3	31.3	24.9	2.17	+	+	+	+
T4-MT625	12.5	26.1	5.9	0.42	−	+	−	−
T5-MT620	7.7	12.4	16.7	0.18	−	+	−	−
T5-MT627	0.7	23.8	16.7	7.26	+	−	−	−
T6-MT549	8.8	3.9	14.9	4.72	+	−	−	+
T6-MT618	16.0	10.5	14.9	9.3	+	−	−	+
T7-MT613	0.7	8.9	17.2	0.24	−	−	−	−
T8-MT617	1.0	24.4	13.2	1.74	−	−	−	−
T9-MT621	8.9	48.1	14.8	1.32	−	−	−	+
T10-MT614	1.4	16.1	5.9	0.21	−	−	−	−
T11-MT618	29.1	31.7	16.5	4.95	+	−	−	−
T12-MT571	9.0	15.6	17.9	8.41	+	−	−	−
T13-MT529	9.6	32.1	135.2	3.90	+	−	−	+
T14-MT550	10.2	26.5	27.0	4.97	+	−	+	−
T16-MT543	26.9	136.8	101.2	3.85	+	+	+	−
Average error	11.9 ± 2.3	29.9 ± 8.1	29.5 ± 9.2					

Estimation errors obtained with the first technique (selective use of MT data) are shown in column labeled "1"; those obtained with the second technique ("blindfold" use of the whole MT dataset available) are shown in column "2," and those obtained from temperature logs alone are shown in column "3." After Spichak et al. (2011).

shows that the error for borehole T16 is two times larger than the average error, whereas the T7 and T9 errors are substantially smaller than the average error. This fact indicates that the geographic factor is only of secondary importance for the temperature estimation, confirming the conclusion made by Spichak (2006).

4.2.3 Effect of the Geology and Hydrological Conditions

The boreholes and MT sites used in this study are located in regions with diverse geological and hydrological conditions. Our experience shows that local geological heterogeneities and extreme hydrological conditions between the temperature estimation point and MT site whose data are used for this assessment (faulting, lateral 2-D or 3-D geological inhomogeneities, cold meteoric water flows) negatively affect the estimation errors. The right side of Table 4.1 indicates the presence ("plus") or absence ("minus") of these factors in each case. Additionally, too big (more than 2 km) distance between the MT site and location of the temperature estimation was considered as a potential risk factor.

As it is seen from Table 4.1, in the cases when the borehole and MT site are located on opposite sides of a tectonic thrust fault (T1-MT550, T5-MT620, T16-MT543) the temperature forecast in the locations of boreholes based on the electrical conductivity profiles at the respective sites results in quite big errors. Similarly, the temperature forecast error greatly increase due to the presence of local zones with a thick (about 200 m) crust layer penetrated by cold water flows that forms an anomalous negative temperature gradient (T13-MT529) (Lesik, 1988).

In this connection it is interesting to compare the results for the pairs T1-MT550 and T14-MT550. The first pair is characterized by presence of the intermediate fault and spacing of 2.17 km, while the spacing in the second pair is 4.97 km. The temperature forecast errors are inversely proportional to the spacing. Similar effect is observed also for the pairs T5-MT620 (presence of the fault) and T5-MT627. The temperature estimation errors carried out in such areas by ANN using only the temperature data are also quite big.

Thus, comparative analysis of the temperature estimation errors in Table 4.1 clearly shows that accuracy of the temperature forecast in the interwell space is controlled by 4 factors: faulting in the space between the place where the temperature profile is estimated and related MT site; spacing between them; meteoric and groundwater flows; and local geological medium heterogeneity (though, the latter factor being less restrictive, if appropriate EM inversion tools are used). We conclude that estimation errors depend on the presence of specific geological features between the temperature estimation point and the EM site whose data are used for the estimation. So, prior knowledge of the geology and hydrological conditions in the region under study can help to correctly locate the EM sites with respect to the points where the temperature is to be predicted and thereby reduce the estimation errors. It is remarkable that elimination from

Table 4.1 of pairs, where at least one of the above mentioned negative factors takes place reduces the average relative error of the temperature estimation from 11.9 to only 1.0%.

4.3 EM TEMPERATURE EXTRAPOLATION IN DEPTH

It is often necessary to estimate the temperature distribution not only in the space between the boreholes, but also at depths that exceed the depth of the drilled wells. Routine MATLAB-based techniques used to this end or linear extrapolation of the temperature gradient often result in erroneous temperature assessment, while existing indirect geothermometers provide the temperature estimations only at some characteristic depths (see Sections 2.1 and 2.2 in Chapter 2). On the contrary, the indirect EM geothermometer could be used for the temperature estimation at arbitrary pre-determined depth (Spichak and Zakharova, 2009a, 2009b) constrained only by the deepness of the EM sounding technique used to this end. We will discuss below the methodological issues related to the temperature extrapolation citing the results obtained by Spichak and Zakharova (2009a, 2009b) for the sedimentary cover (northern Tien Shan, Kyrgyzstan) and geothermal area (Hengill, Iceland).

4.3.1 Sedimentary Cover

The MT and temperature data used for the experiments were collected in the same area of the northern Tien Shan as in the previous studies (see Figure 4.1 from Section 4.2.1).

4.3.1.1 Data and Thermometer Calibration

At the stage of the thermometer calibration the artificial neuronets were taught by correspondence between the temperature values from eight well logs and electrical conductivity profiles (T4-MT625, T5-MT627, T7-MT613, T8-MT617, T9-MT621, T10-MT614, T11-MT618, T12-MT571) estimated from MT data at neighboring sites (Figure 4.2). In order to model the effect of the ratio between the borehole length and the extrapolation depth the whole depth of each well was divided into 10 intervals and the training was carried out successively at 1/10, 2/10, 3/10, ... fractions of the depth. After the training each ANN was tested using the data from remaining part of the temperature profile, the data from which were not used for training.

4.3.1.2 Temperature Extrapolation

Shown in Table 4.2 are the errors of the well temperature estimation at different depths depending on portion δ of temperature profiles and electrical conductivity (from the surface to maximal well depths) used for neuronet calibration. It can be seen from Table 4.2 that for all boreholes the relative testing errors ε decrease monotonically with increasing δ (on average, from 52.39% at

TABLE 4.2 Errors of the temperature extrapolation to the depth in the Tien Shan area depending on the portion δ of the temperature well logs and electrical conductivity profiles at adjacent MT sites used for the neuronet training

δ	T4/ MT625	T5/ MT627	T7/ MT613	T8/ MT617	T9/ MT621	T10/ MT614	T11/ MT618	T12/ MT571
0.1	32.8	67.4	65.5	55.5	33.4	27.3	73.5	63.8
0.2	8.2	36.9	63.3	56.1	34.6	27.3	64.0	24.9
0.3	7.2	12.5	54.2	3.2	37.4	27.5	47.2	24.5
0.4	4.7	5.5	17.2	3.8	1.5	18.3	24.8	8.8
0.5	3.2	1.5	6.8	2.7	6.0	14.0	7.9	4.5
0.6	2.8	2.9	8.4	5.0	2.2	5.0	8.5	5.7
0.7	2.6	2.0	2.3	1.3	1.9	8.3	8.7	4.9
0.8	2.4	4.0	1.8	1.9	2.9	4.1	6.6	1.3
0.9	1.2	1.9	1.4	1.5	0.8	2.3	5.8	1.6

After Spichak and Zakharova (2009a).

$\delta = 0.1–2.44\%$ at $\delta = 0.9$). Yet starting from $\delta = 0.5$ the errors ε of extrapolation in depth become, on average, lower than 10%, although this level for different wells is achieved at different values of δ.

In Figure 4.5 a plot is shown illustrating the dependence of the mean relative error ε of the neuronet prognosis (extrapolation) of temperature in depth (based on electromagnetic data measured at the MT site closest to the well) versus the portion of the electrical conductivity and temperature profiles used for the neuronet training. From this graph one can conclude that, to reach, say, 5–6% level of the estimation error it is quite sufficient to employ for ANN teaching only the temperature and electrical conductivity data for the upper half of the profile. In other words, the use of indirect electromagnetic geothermometer could enable obtaining high-accuracy temperature estimates at depths twice as large as the lengths of the drilled wells for which the temperature data are available.

Figure 4.6 indicates the actual temperature profiles and extrapolated to the lower half depths of all temperature profiles in eight wells. As it is seen from Figure 4.6, only in three of eight cases the predicted curves insignificantly depart from the actual ones, and, moreover, in two cases (T7-MT613 and T12-MT571) the departure is observed only at depths from 2.5 to 4.5 km. Table 4.3 shows the temperature estimation errors for all eight boreholes in comparison with the case if the temperature in the bottom halves of the boreholes is estimated by ANN extrapolation using only the temperature records in the upper halves (provided according to the technique proposed in Spichak (2006)). It is seen that in the former case the average error is 5.8%, while in the latter case it is

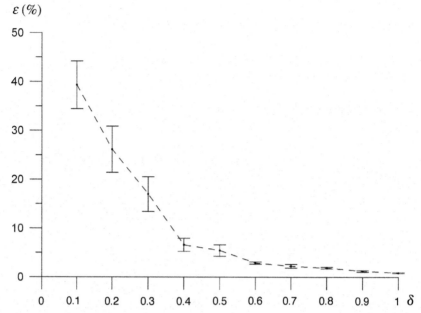

FIGURE 4.5   Dependence of the average relative error ε of the EM temperature extrapolation in the Tien Shan area on the portion δ of the temperature well logs used for the neuronet training (after Spichak and Zakharova, 2009a).

equal to 27.4%. It is worth mentioning in this connection that routine temperature extrapolation based on the MATLAB library results in constant temperature values at depth.

4.3.2   Geothermal Area

4.3.2.1   Data and Thermometer Calibration

For the temperature extrapolation in the Hengill geothermal area Spichak and Zakharova (2009a, 2009b) used, magnetotelluric data measured at eight sites (MT38, MT44, MT46, MT49, MT52, MT53, MT81, MT192) close to the wells T4, T3, T11, T6, T5, T8, T10, and T15, respectively (Figure 4.7). Profiles of apparent conductivity reconstructed from the measured MT data are shown in Figure 4.8 together with the well temperature logs. As it is seen from Figure 4.8, in the majority of sites the temperature monotonically increases with depth reaching values as high as 250–300°C at depths of about 0.8–1.0 km.

At the same time, the apparent electrical conductivity in most MT sites is first increasing with depth and, after its maximum is reached at a depth of 0.5–0.6 km, is then decreasing. According to Oskooi et al. (2005), the presence of an outcropping resistive layer is identified as the typical unaltered porous basalt of the upper crust. This layer is underlain by a highly conductive cap resolved

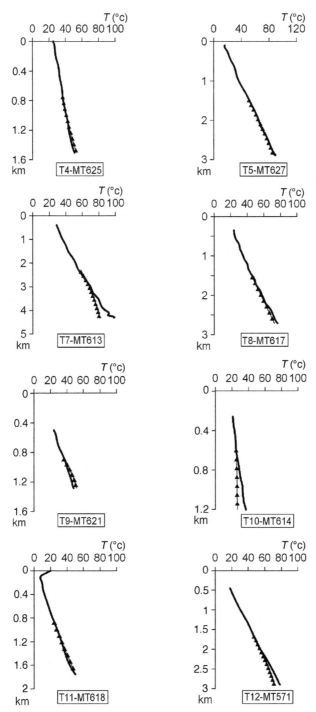

FIGURE 4.6 Well logs (solid lines) and estimated temperature profiles (lines with triangles) in the Tien Shan area obtained using extrapolation on the lower half of the profile by the net trained on the correspondence between the electrical conductivity and temperature at the points belonging to the upper half of the profile (after Spichak and Zakharova, 2009a).

TABLE 4.3 Temperature estimation errors (in per cent) for Tien Shan area depending on the extrapolation technique used: ε corresponds to indirect EM geothermometer, while ε* relates to ANN temperature extrapolation using only the temperature records

Well	MT site	T–MT spacing (km)	ε (%)	ε* (%)
T4	MT625	0.42	3.2	20.0
T5	MT627	0.18	1.5	31.1
T7	MT613	0.24	6.8	28.4
T8	MT617	1.74	2.7	25.3
T9	MT621	1.32	6.0	25.5
T10	MT614	0.21	14.0	22.0
T11	MT618	4.95	7.9	34.0
T12	MT571	8.41	4.5	32.9
Average			5.8 ± 1.3	27.4 ± 1.7

After Spichak and Zakharova (2009a).

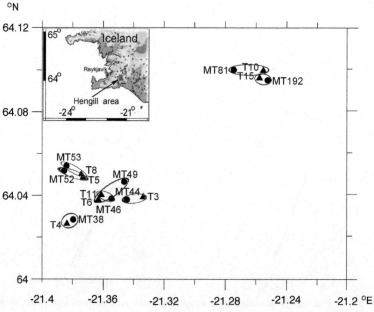

FIGURE 4.7   Location scheme of MT sites (circles) and wells (triangles) for which temperature data are available in the Hengill geothermal area (after Spichak and Zakharova, 2009a, 2009b).

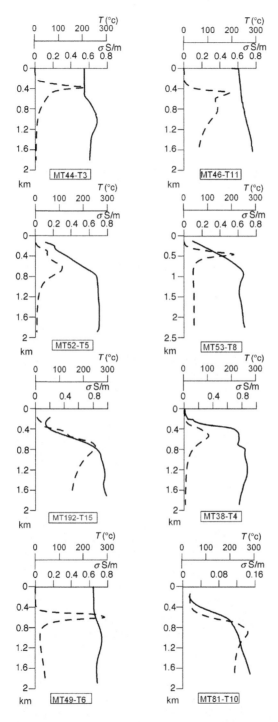

FIGURE 4.8 Temperature well logs (solid lines) and electrical conductivity profiles beneath adjacent MT sites (dashed lines) in the Hengill geothermal area (Spichak and Zakharova, 2009a, 2009b).

as the smectite–zeolite zone. Below this cap a less conductive zone is identified as the epidote–chlorite zone.

Same as in the previous case, for the geothermometer calibration the neuronets were trained by the correspondence between the values of apparent electrical conductivity and temperature within the upper halves of the profiles for each well.

4.3.2.2   Temperature Extrapolation

In Figure 4.9, actual temperature values for eight wells and temperature predictions for the lower half-depths of the profiles are shown. As it is seen from Figure 4.9, the most noticeable discrepancy (particularly at great depths) between the prognostic and actual values is observed for wells T3, T6, T8, and T10. In the case of wells T3 and T6, this could be associated with an anomalous character of temperature changes with depth.

However, also a general reason exists that can explain the divergences in all the four cases. In Table 4.4, the extrapolation errors are shown for all wells, and the distances are indicated between the wells and MT sites providing MT data for the analysis. As it is seen from Table 4.4, these distances are maximal exactly in the cases mentioned above. On the other hand, in areas where they are minimal, in most cases the minimal discrepancies are observed. The correlation coefficient between the extrapolation errors and the spacing between the MT sites and boreholes was found equal to 0.95. This argues for the conclusion that in order to minimize the errors in temperature prognosis at depths exceeding the well lengths, it is necessary to measure electromagnetic data in the closest proximity of the wells.

However, this inference is not supported by the extrapolation results in the Tien Shan area (see Table 4.3). In contrast to the above case no correlation between the extrapolation errors and spacing between MT sites and boreholes was manifested. This could be explained by the inference that the correlation between the electrical conductivity and temperature profiles in this area is governed by their dependence on depth rather than by spacing between the boreholes and MT sites (Zakharova et al., 2007).

It is noteworthy in this connection that the errors of lateral electromagnetic extrapolation of temperature depend rather on the geological heterogeneities of the medium (e.g., faults) than on the distances between the MT sites and wells from which the temperature logs are taken for calibration and could be further diminished if the geology is taken into account during MT survey (see Section 4.2.3).

Comparison with the case if the temperature in the bottom halves of the boreholes is estimated by ANN extrapolation using only the temperature records in the upper halves shows that, as in the Tien Shan area, in the former case the average error is twice less (4.9%) than in the latter one (9.8%) with errors being less in six boreholes from eight.

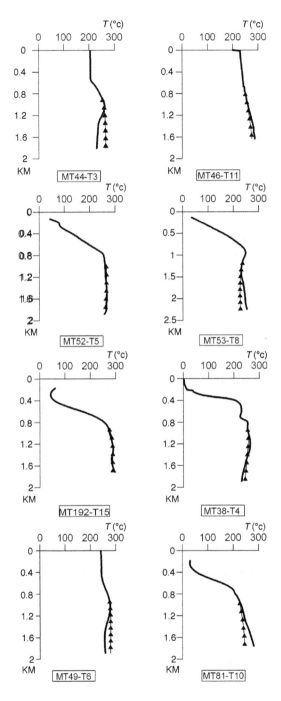

FIGURE 4.9 Well logs (solid lines) and estimated temperature profiles (lines with triangles) obtained using extrapolation on the lower half of the profile by the net trained on the correspondence between the electrical conductivity and temperature at the points belonging to the upper half of the profiles on the Hengill geothermal area (after Spichak and Zakharova, 2009a, b).

TABLE 4.4 Temperature prognosis errors (in per cent) for Hengill area depending on the extrapolation technique used: ε corresponds to indirect EM geothermometer, while ε* relates to ANN temperature extrapolation using only the temperature records

Well	MT site	T–MT spacing (km)	ε (%)	ε* (%)
T3	MT44	0.55	8.9	5.0
T4	MT38	0.27	2.9	8.1
T5	MT52	0.21	2.8	3.5
T6	MT49	0.91	6.0	3.3
T8	MT53	0.42	6.1	25.4
T10	MT81	0.93	7.7	18.8
T11	MT46	0.42	2.7	9.0
T15	MT192	0.30	2.0	4.9
Average			4.9 ± 0.9	9.8 ± 2.7

After Spichak and Zakharova (2009a, 2009b).

4.3.3   Robustness Evaluation

In order to study how the prediction errors depend on the behavior of the electrical conductivity–temperature profiles used for the EM geothermometer calibration and the conductivity profiles used for the temperature extrapolation the following experiments in both studied areas were carried out (Spichak and Zakharova, 2009a, b). Each electrical conductivity and temperature profile was divided into 200-m-thick sections starting from the depth of 150 m. Then for each T–MT pair the neuronets were successively trained on the data of the corresponding intervals. The taught neuronets were then tested on 200-m-thick intervals proximate in depth to the given ones.

Tables 4.5 and 4.6 show the extrapolation errors made in such a way for the Tien Shan area and Hengill geothermal zone, respectively. The analysis of the obtained results indicates that in both cases the average errors of interval prognosis for each well are rather big for subsurface sections (reaching as high values as 87% at T4 well in the Hengill geothermal zone), but farther with depth the errors gradually decrease. In principle, this could be related to MT data distortion by subsurface geological noise (so called static shift effect). On the other hand, for the Hengill geothermal zone the correction of MT curves employing the transient EM data inversion results caused no reduction to the extrapolation errors.

Common reason for such a behavior of errors is presumably a decrease in the vertical inhomogeneity of the medium with depth (more distinct in the Hengill zone and less pronounced in the northern Tien Shan region). In the latter

TABLE 4.5 Temperature estimation errors (%) for Tien Shan area depending on the depth range used for indirect EM geothermometer calibration and testing

	Depth range (km)									
N	Calibration	Testing	T4-MT625	T5-MT627	T7-MT613	T8-MT617	T9-MT621	T10-MT614	T11-MT618	T12-MT571
1	0.15–0.35	0.35–0.55	12.7	15.1				11.3		
2	0.35–0.55	0.55–0.75	12.7	16.1	3.1	14.8	19.3	9.3	28.9	18.8
3	0.55–0.75	0.75–0.95	9	7.5	6	11.6	2.7	14.5	23.8	14.3
4	0.75–0.95	0.95–1.15	7.1	12.6	8.5	11.9	5.2	11.2	23.4	9.3
5	0.95–1.15	1.15–1.35	2.5	11.2	3.1	11.3			13.7	8.9
6	1.15–1.35	1.35–1.55	8.6	8.1	5.9	9.4			12	10.1
7	1.35–1.55	1.55–1.75	8.9	5.7	12.8	10.2			10.9	8.8
8	1.55–1.75	1.75–1.95		11	6.1	9.1			10.3	9.4
9	1.75–1.95	1.95–2.15		6.1	4.6	6.3				8.6
10	1.95–2.15	2.15–2.35		7.4	5	8.7				8.9
11	2.15–2.35	2.35–2.55		7.1	7.3	7.4				5.7
12	2.35–2.55	2.55–2.75		5.1	6.7	5.4				

After Spichak and Zakharova (2009a).

TABLE 4.6 Temperature estimation errors (in per cent) for Hengill area depending on the depth range used for indirect EM geothermometer calibration and testing

N	Depth range (km) Calibration	Testing	T3-MT44	T4-MT38	T5-MT52	T6-MT49	T8-MT53	T11-MT46	T10-MT81	T15-MT192
1	0.15–0.35	0.35–0.55	2.4	87.0	26.9	0.5	21.9	1.9	33.9	57.7
2	0.35–0.55	0.55–0.75	11.0	14.8	13.8	4.1	6.0	2.8	18.1	19.8
3	0.55–0.75	0.75–0.95	3.4	6.0	9.3	6.2	4.5	3.6	2.8	3.4
4	0.75–0.95	0.95–1.15	2.1	9.9	1.6	2.6	2.9	3.9	1.9	0.4
5	0.95–1.15	1.15–1.35	1.9	3.1	1.0	2.7	6.0	4.2	4.6	5.9
6	1.15–1.35	1.35–1.55	2.6	1.3	1.2	4.6	0.7	2.7	6.0	1.1
7	1.35–1.55	1.55–1.75	1.2	5.7	0.4	1.3	2.5	1.2	2.5	1.3

After Spichak and Zakharova (2009a).

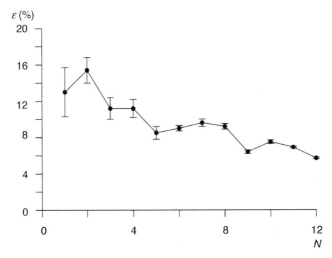

FIGURE 4.10 Temperature estimation errors (in per cent) for Tien Shan area depending on the number of the depth interval used for indirect EM geothermometer calibration and testing (after Spichak and Zakharova, 2009a).

area, with its sophisticated geological structure characterized by essentially 3-D distribution of conductivity, such a decrease of the medium inhomogeneity with depth is slower, which reflects as well in the behavior of the interval errors of the temperature extrapolation based on electrical conductivity data.

Average relative extrapolation errors for all wells in these two regions are shown in Figures 4.10 and 4.11. Regardless the common tendencies of the errors decrease with depth, in the Tien Shan area (Figure 4.10) the curve is less monotonic (due to the above mentioned vertical heterogeneity of the medium) than in the Hengill zone. In the latter case the curves reach their asymptotic values already starting from the depth of about 1 km ($N = 4$ in Figure 4.11) that characterizes the transition to the homogeneous distribution in both the electrical conductivity and temperature for most wells in this geothermal zone (Figure 4.8).

Thus, application of the indirect electromagnetic geothermometer allows high accuracy temperature estimation at depths exceeding the lengths of drilled wells for which temperature data are available. In order to minimize the errors in temperature prognosis at large depths it is necessary to provide electromagnetic sounding in the proximity of the wells. Meanwhile, this recommendation could be soften in the sedimentary cover areas, where the correlation between the electrical conductivity and temperature profiles is governed mainly by their dependence on depth rather than by spacing between the boreholes and EM sites.

On the basis of the obtained results, an important practical recommendation can be proposed: when calibrating an indirect EM geothermometer it is advisable to avoid using the subsurface sections of temperature and electrical conductivity profiles that are characterized by the strongest vertical heterogeneity.

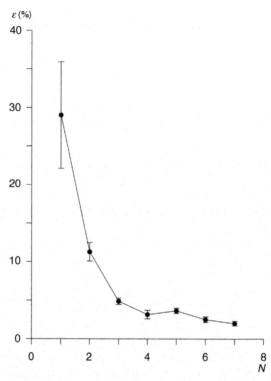

FIGURE 4.11   Temperature estimation errors (in per cent) for Hengill area depending on the number of the depth interval used for indirect EM geothermometer calibration and testing (after Spichak and Zakharova, 2009a, b).

4.4 CONCLUSIONS

The studies carried out using temperature and MT datasets allow us to make important conclusions about a feasibility of estimating the temperature in the earth's interior from electromagnetic sounding data (in particular, magnetotelluric) measured on the surface. In contrary to known indirect geothermometers, which attribute the temperature dependency of the composition of some characteristic hydrothermal components observed at the surface to the supposed depth of their origin, the electromagnetic one provides the temperature distribution in the earth at the absence of manifestations of the geothermal activity on the surface.

It is important mentioning that the temperature estimations by means of indirect EM geothermometer are based on its advance calibration by electrical conductivity – temperature relations in a few wells (in particular, six to eight temperature logs used for calibration of electromagnetic data turn to be sufficient for ensuring a 12% accuracy of temperature prediction in the given point). The indirect temperature estimations do not depend explicitly on their nature possibly affected by alteration mineralogy or other factors and result in more accurate temperature estimations than those obtained using any interpolation or extrapolation of the temperature well logs.

It is shown that the interwell temperature estimation errors are controlled by four factors: faulting in the space between the place where the temperature profile is estimated and related MT site, distance between them, meteoric and groundwater flows and lateral geological inhomogeneity (though, the latter factor being less restrictive, if appropriate EM inversion tools are used). Therefore, prior knowledge of the geology and hydrological conditions in the region under study can help to correctly locate the MT sites with respect to the points where the temperature is to be predicted and thereby reduce the estimation errors (in particular, from 11.9 to 1.0% in the Tien Shan case).

Application of the indirect electromagnetic geothermometer allows high accuracy temperature estimation at depths exceeding the lengths of drilled wells for which temperature data are available. For example, when extrapolating to a depth twice as large as the well depth the relative error is 5–6%, and if the extrapolation depth is three times as large, the error is about 20%. This result makes it possible to increase significantly the deepness of the indirect temperature estimation in the earth's interior (in particular, for geothermal exploration) based on the available temperature logs in wells.

Thus, usage of the indirect EM geothermometer enables, first, to estimate the subsurface temperature distribution (especially, in cases when the number of temperature logs available is insufficient) and, second, to perform temperature estimations in extrapolation mode (both in horizontal and vertical axes).

In the next chapters we will consider some case studies, where application of the indirect EM geothermometer enabled to forecast the temperature at large depths, which, in turn, helped to answer to important questions related to geothermal exploration.

REFERENCES

Chiles, J.-P., Delfiner, P., 1999. Geostatistics: modeling spatial uncertainty. John Willey and Sons.

Duchkov, A.D., Schwartzman, Yu.G., Sokolova, L.S., 2001. Tien-Shan deep heat flow: developments and problems. Russian Geol. Geophys. 42 (10), 1512–1529.

Haykin, S., 1999. Neural networks: a comprehensive foundation, second ed. Prentice Hall.

Lesik S O.M., 1988. Deep structure of the frunze prediction research area, Cand. Sc. (Geol.-Mineral.) Th., Frunze. Institute Seismol. Acad. Sci. Kyrgyz SSR, (in Russian).

Oskooi, B., Pedersen, L.B., Smirnov, M., Árnason, K., Eysteinsson, H., Manzella, A., 2005. The deep geothermal structure of the Mid-Atlantic Ridge deduced from MT data in SW Iceland. Phys. Earth Planet. Inter. 150, 183–195.

Park, S.K., Livelybrooks, D.W., 1989. Quantitative interpretation of rotationally invariant parameters in magnetotellurics. Geophysics 54 (11), 1483–1490.

Spichak, V.V., 2006. Estimating temperature distributions in geothermal areas using a neuronet approach. Geothermics 35, 181–197.

Spichak, V.V. (Ed.), 2007. Electromagnetic sounding of the earth's interior. Elsevier, Amsterdam.

Spichak, V.V., 2011. Application of ANN based techniques in EM induction studies. Petrovský, E., Herrero-Bervera, E., Harinarayana, T., Ivers, D. (Eds.), The earth's magnetic interior, IAGA special Sopron book series, 1, Springer, pp. 19–30.

Spichak, V.V., Zakharova, O., 2009a. The application of an indirect electromagnetic geothermometer to temperature extrapolation in depth. Geophys. Prosp. 57, 653–664.

Spichak, V.V., Zakharova, O., 2009b. Electromagnetic temperature extrapolation in depth in the Hengill geothermal area, Iceland (Expanded Abstr). XXXIV Workshop on Geothermal Reservoir Engineering, Stanford, USA.

Spichak, V.V., Zakharova, O.K., 2010. Indirect electromagnetic geothermometer: methodology and case study (Expanded Abst). World Geothermal Congress, Bali, Indonesia.

Spichak, V.V., Zakharova, O.K., 2011. Indirect electromagnetic geothermometer – a novel approach to the temperature estimation in geothermal areas. Trans. Geothermal Resource Council 35, 1759–1766.

Spichak, V.V., Zakharova, O.K., Rybin, A.K., 2007. On the possibility of realization of contact-free electromagnetic geothermometer. Dokl. Russian Acad. Sci. 417A (9), 1370–1374.

Spichak, V.V., Zakharova, O.K., Rybin, A.K., 2011. Methodology of the indirect temperature estimation basing on magnetotelluruc data: northern Tien Shan case study. J. Appl. Geophys. 73, 164–173.

Zakharova, O.K., Spichak, V.V., Rybin, A.K., Batalev, V.Yu., Goidina, A.G., 2007. Estimation of the correlation between the magnetotelluric and geothermal data in the Bishkek geodynamic research area. Izvestya Phys. Solid Earth 43 (4), 297–303.

Part II

Case Studies

Chapter 5

Estimation of the Deep Temperature Distribution in the Chu Depression (Northern Tien Shan)

Chapter Outline

The indirect electromagnetic geothermometer was used for building of the deep temperature cross-section in the Chu depression area (northern Tien Shan) based on the magnetotelluric sounding data and available temperature logs (Spichak et al., 2011; Spichak and Zakharova, 2010, 2013). Below we will consider the main results of the studies following these publications.

5.1 GEOLOGICAL SETTING AND THE REGIME OF THE UNDERGROUND WATERS

The region under study (Bishkek geodynamical ground) is a part of the Chu depression (see Figure 4.1 from the Chapter 4). Among all the studied Kyrgyz depressions, the Chu depression is studied best due to the large volume of test drilling, geophysical works, and numerous targeted studies (Mikolaichuk et al., 2003; Laverov, 2005). The Chu depression is an asymmetric structure with a flat northern wing and a steep southern one. Within the depression, the Near Chu monocline, East Chu flexure-rupture zone and Pre-Kyrgyz flexure are distinguished. The Bishkek geodynamical ground where the MT and temperature measurements were carried out is located in the west of the northern Tien Shan area.

The groundwater regime is controlled by the regions of surface runoff, the decrement and shallow occurrence of groundwater, and self-outflow of artesian

Electromagnetic Geothermometry. http://dx.doi.org/10.1016/B978-0-12-802210-8.00005-8

water. Groundwater flow of the pore and pore–fracture circulation types prevails in marginal parts of the Chu basin. Pore water filling in the sequence of Quaternary deposits is the most widespread. These 10–800-m-thick deposits include intense low temperature flows of free level groundwater within the alluvial bench and are fed through the percolation of surface and river waters (Zakharova et al., 2007).

5.2   HEAT FLOW AND TEMPERATURE LOGS

In the Tien Shan folded area, the heat flow and temperature distributions were studied on a global scale using averaged temperature data. Results of these studies are represented as heat flow maps at depths down to about 200 km (Yudakhin, 1983). Valuable contribution to the study of geothermal conditions in the studied region was made by works carried out by Shakirov et al. (1978), Schwartzmann (1984, 1989), and Pogozhev (1993). These works revealed an extremely complicated structure of the heat field within the Chu depression caused by orographic, hydrographic, climate, and geological conditions.

In 1980s, the regional studies of the heat flow were carried out in the submountain and depression regions of Tien Shan (Figure 5.1). In the western part of the northern Tien Shan including the Chu depression, low heat flow not higher than 50 mW/m^2 is observed. The average error of the heat flow estimation is 10–15 mW/m^2 and sometimes even higher. Such data allow only reliable identification of strong heat flow anomalies exceeding 20–30 mW/m^2 (Yudakhin, 1983).

Unfortunately, measurements of the heat flow were conducted very irregularly. In order to mitigate the influence of different exogenous factors such as seasonal variations in well temperature, sedimentation, and climate variations during recent geological epochs, the well geothermal data only from the depths 1000–1500 m were used in the calculations of the heat flow.

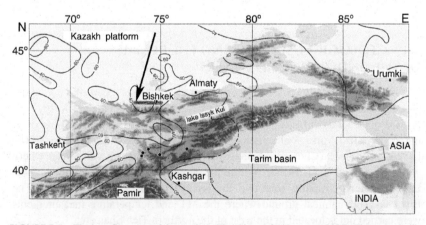

FIGURE 5.1   The heat flow map of the northern Tien Shan (after Laverov, 2005). The arrow indicates the location of the temperature cross-section built using MT data by Spichak and Zakharova (2013).

Within this interval all the analyzed deep wells penetrated the Neocene sediments composed predominantly of sand–clay aleurolites. In order to eliminate the distortions caused by convective heat outflow in the deep fault zones, the wells were sunk within tectonically quiet regions such as the central parts of intermountain depressions. Average temperature gradient revealed by indirect chemical and isotopic geothermometers equals to 2.5 °C/100 m (Yudakhin, 1983), while the characteristic thermal conductivity is around 1.76 W/mK (Duchkov et al., 2001).

In most boreholes of this area the temperature logs show a nonlinear behavior and are characterized by either positive or negative temperature gradients (see Figure 4.2 from the Chapter 4). It is interesting to note that in some logs a negative geothermal gradient is observed down to depths about 200 m (Wells T11, T13) and even up to 400 m (Wells T1, T6). Such a negative temperature gradient is observed everywhere in the Chu depression, moving to the north from the Tien Shan orogene, where cool flows are observed at depths of approximately 200 m (Pogozhev, 1993). The presence of such an effect is explained by, first of all, predominance of permeable rocks in lithological cross-section and by the fact that the wells with anomalous negative temperature gradient are located in the front of the water seepage zone. Main supplying sources for these waters are the melt waters of glaciers. Figure 5.2 indicates three temperature logs having the negative temperature gradients typical for the study area.

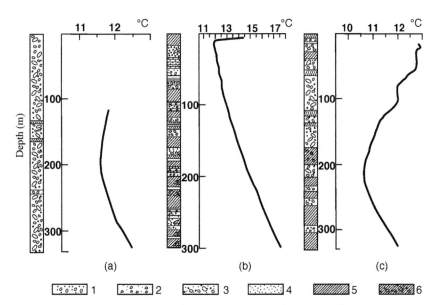

FIGURE 5.2 Temperature well logs and appropriate lithological columns in Chu depression Quaternary sediments of the northern Tien Shan (after Pogozhev, 1993): 1 – sandy-gravel sediment; 2 – pebbles; 3 – boulder-pebble sediments with sandy-gravel loading; 4 – sand; 5 – loam; 6 – loam with pebble and sand.

5.3   EM SOUNDINGS

A number of EM soundings in the region of Bishkek geodynamic ground were carried out (see, for instance, Trapeznikov et al., 1997; Berdichevsky et al., 2001; Rybin et al., 2008). As to the electrical properties of rocks, the cross-section of the Chu depression was studied best by parametric soundings in a series of wells. In accordance with Yudakhin (1983), specific resistivity of rocks within the Bishkek geodynamical ground varies within wide limits. On the whole, the orogenic floors of sedimentary formations are characterized by high resistivities (200 to a few thousands of Ωm). Lower resistivities (up to 20 Ωm) are typical of the cratonic rocks that form the lower part of the Chu depression.

Resistivity of the sedimentary formations of the Chu depression decreases with depth from 250 to 300 Ωm in the upper layer of boulder-bed sediments down to 10 Ωm in the Kyrgyz assise formations and reaches its minimum (3–10 Ωm) in the lower part of sedimentary complex formed by red clays and loams (often salinized and gypsum). These gypsum formations considerably affect the electrical resistivity – the higher is the salt and gypsum content, the lower is the rock resistivity.

Typical feature of geoelectrical structure of the Chu depression is the presence of epi-reference low-resistivity horizon identified as argillo-arenaceous Paleogene–Neogene sediments (Yudakhin, 1983). Variations in the electrical resistivity within this horizon are caused by facial changes in these formations, that is, these variations are indirectly indicative of the conditions of sedimentation and structural specificity existed at that time. Within the Chu monocline in the epi-reference geoelectrical horizon a tendency is apparent of resistivity decrease from the north–east to the south–west. Besides, as the experience shows of geologists involved for many years in the study of the region (Lesik, 1988; Makarov, 1990; Makeev, 2004), in the eastern part of the Chu depression the behavior of electrical conductivity in the vicinity of specific wells can be explained by geological peculiarities at corresponding depths.

5.4   ESTIMATION OF THE ELECTRICAL CONDUCTIVITY AND TEMPERATURE CORRELATION

Zakharova et al. (2007) have carried out a study aimed at estimation of the correlation between the electrical conductivity determined from MT sounding and temperature data collected in the northern Tuen Shan area. To this end 15 borehole–MT site pairs were selected (Figure 4.1). The depth dependences of the electrical conductivity and temperatures from adjacent boreholes are plotted in Figure 4.2. The boreholes and MT sites are located in regions with diverse geological settings and at different distances from each other, which allowed to estimate the influence of these factors on the correlation of the electrical conductivity with temperature. The borehole–MT site distance varied from 0.18 to 1.74 km for six pairs and exceeded 4 km for the other seven pairs.

TABLE 5.1 Correlation ratios between the electrical conductivity and temperature values measured in adjacent wells in the Chu depression area

Number of the borehole	Number of MT site	Spacing (km)	Correlation ratio
T1	MT550	2.17	0.41
T4	MT625	0.42	−0.33
T5	MT627	0.18	0.98
T5	MT620	7.26	0.62
T6	MT549	4.72	0.90
T6	MT618	9.3	0.98
T7	MT613	0.24	0.98
T8	MT617	1.74	0.92
T9	MT621	1.32	0.90
T10	MT614	0.21	0.25
T11	MT618	4.95	0.81
T12	MT571	8.41	0.94
T13	MT529	3.90	−0.73
T14	MT550	4.97	0.79
T16	MT543	3.85	−0.01

After Zakharova et al., 2007.

The correlation ratios between temperatures measured in boreholes and electrical conductivity values at the same depths at the neighboring MT sites are presented in the Table 5.1. The resulting values of correlation ratios reflect the diversity of electrical conductivity versus temperature dependences associated with the complex geological and hydrological conditions of the region under study. Below we consider the reasons of such correlation (or its absence) following (Zakharova et al., 2007).

To estimate the influence of the borehole–MT site spacing on the electrical conductivity correlation with temperature, analyzed were not only the data from the MT sites located near boreholes T5 and T6 but also the MT data from sites MT620 and MT618, respectively, located at substantially greater distances from the boreholes. At the same time, electrical conductivity profiles from sites MT618 and MT550 were analyzed together with the temperature profiles measured in boreholes T6, T11 and T11, T14, respectively.

The low correlation between the temperatures measured in boreholes T1 and T16 and the electrical conductivity at the respective sites MT550 and MT543 can be due to the fact that, in both pairs, the borehole and the MT sites are located on opposite sides of a deep fault nearly reaching the earth's surface.

Siltstone–gypsum salt-bearing deposits that occur near borehole T4 (site MT625) at depths of 0.2–0.4 km form conducting horizons responsible for an increase in the electrical conductivity at a depth of about 0.2 km and its subsequent abrupt decrease below a depth of 0.4 km. This explains the negative correlation between the temperature measured in borehole T4 and the electrical conductivity values at the very close site MT625 (Makarov, 1990; Trapeznikov et al., 1997; Mikolaichuk et al., 2003).

Borehole T5 and site MT620 are located on opposite sides of a tectonic thrust-type fault, which yields a low correlation, whereas T5 and MT627 are located on the same side of the fault (Mikolaichuk et al., 2003), and a high correlation is observed here. Moreover, both the electrical conductivity profile at site MT627 and the thermogram of borehole T5 have positive gradients, which is normal for the thermogram and, in the case of the electrical conductivity, is accounted for by the replacement of high-resistivity coarse clastic rocks by relatively conducting alternating sandy–clayey rocks.

The temperature in borehole T6 is determined only in the depth interval 0.2–0.4 km and, accordingly, the correlation ratio with the electrical conductivity at site MT549 is determined in the same interval, which results in a high correlation, as in the previous case.

Site MT617, the nearest to borehole T8, is characterized by a significant rise in the electrical conductivity at depths of about 1.5–2.5 km, which is apparently associated with the replacement of Upper Neogene coarse clastic sediments (depths of about 0.4–1.2 km) by Upper Paleogene–Lower Neogene sandy–clayey deposits of the Kyrgyz Formation containing magnetic rocks (Lesik, 1988; Makeev, 2004). In above cases, we observe a very high T–MT correlation, which reflects geological features at the considered sites.

Site MT621, located near borehole T9, is characterized by a significant drop in the electrical conductivity at depths of 0.5–0.8 km due to the transition from conducting clays and clayey–sandy rocks of the Chu formation to low-conductivity gravelly–sandy sediments. A transition to clays is observed below a level of 0.6 km. In this case, the high correlation is likely due to the fact that the temperature was measured only from a depth of 0.5 km, at which the temperature and electrical conductivity start to increase. On the whole, the Paleogene–Neogene deposits form a regional confining layer with a slower or stagnant pattern of groundwater motion (Velikhov and Zeigarnik, 1993; Makeev, 2004).

A similar pattern of the electrical conductivity distribution is observed in the areas of boreholes T10 and T11 (sites MT614 and MT618, respectively). In the case of the T10–MT614 pair, an anomalously abrupt drop in the electrical conductivity to depths of 0.8 km, along with a normal temperature increase with depth, leads to a very low correlation coefficient. A good correlation observed in the T11–MT618 pair can be explained, on the one hand, by a much smaller gradient of the electrical conductivity decreasing at depths from 0.35 to 0.8 km and, on the other hand, by a much larger (compared to the T10–MT614 pair) depth limiting the available temperature measurements.

In borehole T13 with adjacent site MT529, the temperature was measured from the earth's surface to a depth of only 0.35 km. These measurements fall into the interval where the electrical conductivity profile practically intersects the thermogram, and the latter has a negative temperature gradient caused by the presence of cold temperature flows. The low electrical conductivity in this interval is due to the presence of Quaternary boulder–pebble rocks composing the alluvial fans of ancient rivers (Lesik, 1988; Makarov, 1990; Makeev, 2004). In the area of borehole T14 and site MT550, a good correlation is likely due to a too short interval of temperature measurements (only 250 m) rather than to any real factors.

Thus, the analysis of the correlation between the electrical conductivity and temperature up to the depths of 3–4 km has led to the following conclusions. First, the correlation ratios do not depend on the distance between a borehole and the nearest MT site; second, a good correlation between the electrical conductivity and temperature in the majority of cases is due to behavior of the study parameters with depth that is normal for the layered sedimentary cover; finally, the lowest correlation ratios are associated with specific features of the geological structure between an MT site and adjacent borehole (presence of tectonic fractures/faults, local heterogeneities affecting the electrical conductivity behavior, lateral low temperature flows). So, a join analysis of the MT and geothermal data could become substantially more effective if geological features of the territory under study are known in advance.

5.5   BUILDING OF THE DEEP TEMPERATURE CROSS-SECTION

5.5.1   MT Data and Dimensionality Analysis

The locations of MT sites used in the studies carried out in (Spichak and Zakharova, 2010; Spichak et al., 2011) are shown in Figure 4.1. The MT data were collected using two Phoenix MTU-5 stations (in the frequency range $5 \cdot 10^{-4}$ to 300 Hz) and two MT-PIK measuring units (in the frequency range $0.28 \cdot 10^{-3}$ to 16 Hz). The MT-PIK data were processed by means of a narrow-band filtering algorithm (Berdichevsky et al., 2001) and the RRRMT-8-1987 code developed by A. Chave. The MTU-5 data were processed with the use of the SSMT2000 software, standard for this instrumentation, in the local and the remote reference modes. In addition, the CORRECTOR code, developed by the North–West Ltd. was used for suppressing industrial noise and smoothing the transfer functions in the low frequency range.

Before using the EM data for indirect temperature estimation, their dimensionality analysis was carried out (Spichak et al., 2011). As is known, lateral geological irregularities in the medium manifest themselves in the frequency dependencies of the so-called dimensionality indicators of the medium, which are equal to zero in a locally homogeneous medium and the more differing from zero the more laterally inhomogeneous the medium is.

This estimation was obtained by means of a joint analysis of a 2-D / 3-D inhomogeneity MT parameter (Berdichevsky et al., 1998) together with the three-dimensionality indicator "skew" (Swift, 1967). According to the former publication the parameter of lateral geoelectrical inhomogeneity, N, could be defined as

$$N = \left| \frac{Z_p^+ - Z_p^-}{Z_p^+ + Z_p^-} \right|, \tag{5.1}$$

where Z_p^+ and Z_p^- are the principal values of the impedance tensor determined by means of the Eggers eigenstate formulation (Eggers, 1982):

$$Z_p^{\pm} = \frac{Z_1}{2} \pm \sqrt{\frac{Z_1^2}{4} - Z_{\text{eff}}^2},$$
$$Z_1 = Z_{xy} - Z_{yx}, \quad Z_{\text{eff}} = \sqrt{Z_{xx}Z_{yy} - Z_{xy}Z_{yx}}, \tag{5.2}$$

Z_{xx}, Z_{xy}, Z_{yx}, Z_{yy} are the elements of the impedance tensor.

The Swift's indicator of three-dimensionality is determined according the formula:

$$S = \left| \frac{Z_{xx} + Z_{yy}}{Z_{xy} - Z_{yx}} \right|. \tag{5.3}$$

Figure 5.3 shows frequency dependencies of N and S parameters. The common feature in the N behavior for all sites is a dramatic increase of N values in the right-hand parts of the curves: from $N = 0.1$–0.2 (for periods shorter than 1 s) to $N = 0.6$–0.8 (for periods longer than 1 s). In view of the electrical conductivity values revealed from MT data, this can be indicative of a strong lateral inhomogeneity of the medium developing in most cases at depths larger than approximately 5 km. Similar tendency is seen also in the behavior of the three-dimensionality indicator, S. On the other hand, at MT543 and MT550 sites enhanced values of the both indicators are observed even at short periods of MT field ($N > 0.2$–0.3 and $S > 0.3$). Thus, for the depth range up to approximately 5 km it was possible to use 1-D electrical resistivity profiles (except sites MT543 and MT550).

5.5.2 Application of the Indirect EM Geothermometer

The deep temperature model along the profile crossing the MT sites and boreholes shown in the Figure 4.1 and Figure 5.1 was built in two stages. At the calibration stage the ANNs were taught by the correspondence between the electrical conductivity and temperature profiles from six T-MT pairs (T5-MT627, T7-MT613, T8-MT617, T9-MT621, T10-MT614, T11-MT618) selected in accordance with four criteria formulated in the Section 4.2.3 (Figure 5.4). It is

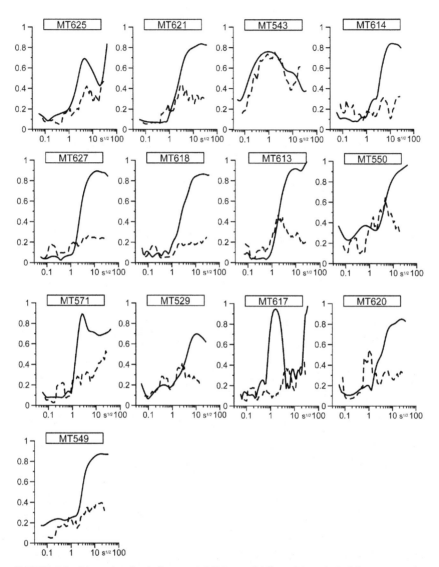

FIGURE 5.3 Dimensionality indicators (2-D/3-D – solid lines; 3-D – dashed lines) versus the square root of the MT field period (in the logarithmic scale) (after Spichak et al., 2011).

worth mentioning in this connection that despite of the electrical conductivity–temperature correlation ratios for these pairs demonstrate maximal values (Table 5.1), this factor does not directly affect the results of the artificial neural network-based temperature reconstruction using the EM geothermometer (Zakharova et al., 2007).

At the second stage the ANNs taught by the electrical conductivity and temperature data for selected pairs were used for the temperature extrapolation

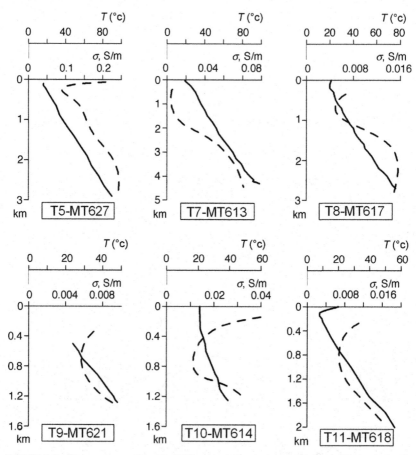

FIGURE 5.4   Temperature well logs (solid lines) and electrical conductivity profiles beneath adjacent MT sites (dashed lines) selected for the EM thermometer calibration in the northern Tien Shan area (Spichak and Zakharova, 2009).

beneath the appropriate boreholes up to the depth of 5 km. Generally speaking, in order to reconstruct the temperature model using the EM geothermometer it is necessary, first, to build the electrical conductivity/resistivity model of the study area. However, in this specific case the distances between the available MT sites along the considered profile were too big (tens of km) for building of a reasonable electrical conductivity model in a local scale. So, in order to reconstruct the temperature cross-section along the E–W profile between the end sites MT613 (adjacent to the borehole T7) and MT621 (adjacent to the borehole T9) an artificial neural network-based interpolation of the deep temperature profiles obtained at the previous stage was carried out. Figure 5.5 indicates the resulting temperature model up to the depth 5 km along the E–W profile shown in Figure 5.1.

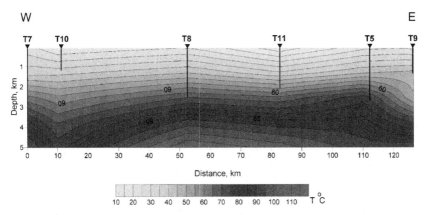

FIGURE 5.5 Temperature cross-section along the E–W profile indicated in Figure 5.1 (after Spichak and Zakharova, 2013).

Analysis of the temperature cross-section shown in Figure 5.5 enables to draw a conclusion about existence of temperature anomalies at depths of 4–5 km beneath the boreholes T7, T8, and T5. According to the regional 3-D resistivity model of the study area (Spichak et al., 2006) shown in Figure 5.6, the western part of the temperature cross-section corresponds to the low-resistivity anomaly deepening beneath the borehole T7 in the NW direction up to the depths 4–5 km. It correlates with the increase of the heat flow at the surface from 40 to 60 mW/m² in the same direction (Figure 5.1). High-temperature and low-resistivity anomalies detected in this area may be related

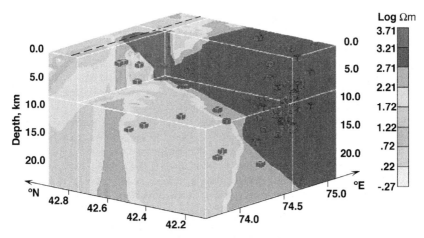

FIGURE 5.6 Regional resistivity model of the northern Tien Shan crustal area (after Spichak et al., 2006). Red elementary boxes mark zones of maximal correlation between bulk resistivity and density of the earthquake hypocenters, dashed line indicates the projection of the temperature cross-section built using EM geothermometer.

with convection of the hot fluids along active faults crossing this area in the E–W direction (Laverov, 2005).

REFERENCES

Berdichevsky, M.N., Dmitriev, V.I., Pozdnjakova, E.E., 1998. On two-dimensional interpretation of magnetotelluric soundings. Geophys. J. Int. 133, 585–606.

Berdichevsky, M.N., Dmitriev, V.I., Vanyan, L.L., 2001. Deep geoelectric studies of the earth's crust and upper mantle and some results obtained in the mountains of Tien Shan. Deep electromagnetic soundings of the mantle around the Teisseyre–Tornquist Zone (Expanded Abstr.). NATO Advanced Research Workshop Belsk Duzy, Poland, pp. 12–13.

Duchkov, A.D., Schwartzmann, Yu.G., Sokolova, L.S., 2001. The deep heat flow of Tien-Shan: advances and challenges. Russian Geol. Geophys. 42 (10), 1512–1529.

Eggers, D.E., 1982. An eigenstate formulation of the magnetotelluric impedance tensor. Geophysics 47, 1204–1214.

Laverov, N.P. (Ed.), 2005. Contemporary geodynamics of the regions of intracontinental collision orogenesis (Central Asia). Scientific World, Moscow (in Russian).

Lesik, O.M., 1988. Deep structure of the Frunze prognostic ground. Seismology Institute of the Kyrgyz Academy of Sciences, Ilim Publ, Frunze (PhD thesis, in Russian).

Makarov, V.I., 1990. The newest orogenes, their structure and geodynamics (Abstr. dissertation). GIN RAS, Moscow (in Russian).

Makeev, V.P., 2004. Studies of structure, composition, hydrodynamic conditions and collecting properties of Chujsko-Issykkul Phanerozoic rocks. 2000-2004 Rpt. of the Kyrgyz expedition of the State Agency on Geology and Mineral ReservesState Agency on Geology and Mineral Reserves, Bishkek (in Russian).

Mikolaichuk, A.V., Sobel, E., Gubrenko, M.V., Lobanenko, A.N., 2003. Structural evolution of the northern periphery of the Tien-Shan orogene. Kyrgyz NAS Trans 4, pp. 50–58 (in Russian).

Pogozhev, I.P., 1993. Geothermal studies in the region of Chu depression. In: Geothermics of seismic and aseismic zones. Nauka, Moscow, pp. 261–268 (in Russian).

Rybin, A.K., Spichak, V.V., Batalev, V.Yu., Bataleva, E.A., Matyukov, V.E., 2008. Array magnetotelluric soundings in the seismogenic area of the northern Tien Shan. Russian Geol. Geophys. 5, 445–460.

Schwartzmann, Yu.G., 1984. Heat fields of Kyrgyz Tien Shan. Geological and geophysical exploration of seismo-hazardous regionsIlim Publ, Frunze, pp.100–120 (in Russian).

Schwartzmann, Yu.G., 1989. Geothermal regime of the Tien-Shan seismic layer. Seismicity of the Tien-ShanIlim Publ, Frunze, pp. 217–230 (in Russian).

Shakirov, E.Sh., Schwartzmann, Yu.G., Palamarchuk, V.K., 1978. Specific features of the deep structure of the North Kyrgyzia and their relation to seismicity and geothermics. Geological and geophysical structure and seismicity of KyrgyziaIlim Publ, Kyrgyz AS, pp. 30–37 (in Russian).

Spichak, V.V., Rybin, A., Batalev, V., Sizov, Yu., Zakharova, O., Goidina, A., 2006. Application of ANN techniques to combined analysis of magnetotelluric and other geophysical data in the northern Tien Shan crustal area (Expanded Abstr.). 18th IAGA WG 1. 2 Workshop on Electromagnetic Induction in the Earth. El Vendrell, Spain.

Spichak, V.V., Zakharova, O., 2009. The application of an indirect electromagnetic geothermometer to temperature extrapolation in depth. Geophys. Prosp. 57, 653–664.

Spichak, V.V., Zakharova, O., 2010. Indirect EM temperature estimation in the northern Tien Shan (Expanded Abstr). 20th IAGA WG 1. 2 Workshop on Electromagnetic Induction in the Earth. Giza, Egypt.

Spichak, V.V., Zakharova, O.A., 2013. The electromagnetic geothermometer. Scientific World, Moscow (in Russian).

Spichak, V.V., Zakharova, O.K., Rybin, A.K., 2011. Methodology of the indirect temperature estimation basing on magnetotelluruc data: northern Tien Shan case study. J. Appl. Geophys. 73, 164–173.

Swift, C.M., 1967. A magnetotelluric investigation of an electrical conductivity anomaly in the south western United States. PhD Thesis, MII, Cambridge, MA.

Trapeznikov, Yu. A., Andreeva, E.V., Batalev, V.Yu., 1997. Magnetotelluric Soundings in the Kyrgyz Tien Shan. Izvestiya Phys. Solid Earth 33, 1–17.

Velikhov, E.P., Zeigarnik, V.A. (Eds.), 1993. Implications of geodynamic processes for geophysical fields. Nauka Publ, Moscow, 158 pp. (in Russian).

Yudakhin, F.N., 1983. Geophysical fields, deep structure and seismicity of the Tien Shan. Ilim Publ, Frunze (in Russian).

Zakharova, O.K., Spichak, V.V., Rybin, A.K., Batalev, V.Yu., Goidina, A.G., 2007. Estimation of the correlation between the magnetotelluric and geothermal data in the Bishkek geodynamic research area. Izvestya Phys. Solid Earth 43 (4), 297–303.

Chapter 6

Gaseous Versus Aqueous Fluids: Travale (Italy) Case Study

6.1 INTRODUCTION

Geothermal deposits are perfect objects for their study by electromagnetic methods since they cause significant variations in electric resistivity of the geological medium. However, in contrast to the high temperatures, which are a necessary condition for searching for a geothermal reservoir, the low or high electrical resistivity itself is not indicative neither of the presence of the reservoir nor of the character of the fluid (gaseous or aqueous) circulating in the hydrothermal system.

Despite this, in many studies devoted to identifying the geothermal reservoirs from the ground-based geophysical (including electromagnetic) data (see Chapter 1), interpretation of the deep resistivity sections is mainly focused on revealing the highly conductive areas, which are then considered as the geothermal reservoirs or the pathways of circulation of a liquid geothermal fluid. In order to reduce the uncertainty and to conduct the interpretation of the geophysical data in the terms of geothermal application, it would be desirable to analyze the geoelectrical cross-sections together with the distributions of temperature. However, the difficulty in reconstructing the temperature sections is associated with the fact that temperature logs can only provide temperature distributions up to a depth of 2–3 km, while the heat flow modeling yields only the regional-scale reconstructions (Khutorskoi et al., 2008; Limberger and Van Wees, 2013), which are lacking the necessary accuracy.

Meanwhile, indirect electromagnetic geothermometer enables to reconstruct the 2-D and 3-D temperature models with accuracy sufficient for further

Electromagnetic Geothermometry. http://dx.doi.org/10.1016/B978-0-12-802210-8.00006-X
117

analysis in geothermal terms. In this chapter, by the example of the Travale geothermal area, we explore the possibility of estimating the type of the heat carrier (liquid or gaseous fluid) and searching for geothermal reservoirs by joint analysis of the resistivity cross-section and temperature distribution yielded by this approach following (Spichak and Zakharova, 2014).

6.2 GEOLOGICAL SETTING

Travale is a part of the world famous Larderello geothermal region in Italy. Its location is shown in Figure 6.1a. Here, two geothermal basins characterized by very high temperatures and high vapor release are exploited. The shallow basin is located at a depth of the cataclastic horizons of carbonate evaporite deposits belonging to the Toskana complex. The deeper reservoir is larger; it is hosted by the fractured metamorphosed rocks at a depth below 2 km (Barelli et al., 2000).

The following tectonostratigraphic complexes (Figure 6.1b) are distinguished in the area of the study (Brogi et al., 2003; Bellani et al., 2004):

1. the Miocene–Pliocene and Quaternary sediments filling the vast tectonic depressions;
2. the Ligurian complex that includes Jurassic ophiolites, the overlying Jurassic–Cretaceous sediments, and Cretaceous–Oligocene flysch formations;
3. the Tuscany complex composed of the sedimentary rocks, Cretaceous and Oligocene evaporite and carbonate rocks up to the Late Oligocene–Early Miocene turbidites;

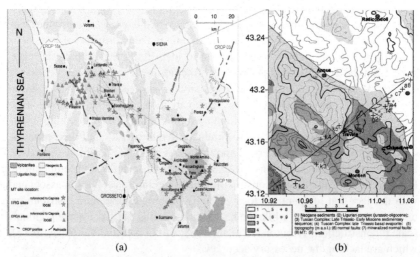

(a)   (b)

FIGURE 6.1   (a) Simplified geological map of the Travale area; (b) map showing locations of MT sites (crosses), profile A–A′ and boreholes (red dots) (after Brogi et al., 2003; Manzella et al., 2006).

4. the underlying strata revealed by the geothermal drilling. These rocks include two units. The upper unit is related to the Monticiano–Roccastrada formation, which is mainly composed of the Triassic quartzites and phyllites (the Verrucano group), Paleozoic phyllites and mica schists. The lower unit corresponds to the gneiss complex. The rocks composing the upper unit bear the footprints of the Apennine orogeny, whereas the gneiss complex does not show any signs of such transformation. This probably indicates that these rocks were part of the foreland crustal area ahead of the Apennine orogen.

After the stages of convergence and collision (in the Late Cretaceous–Early Miocene), which determined the further structural development of the northern Apennines, the Larderello region experienced three episodes of tectonic extension. The first and second events occurred in the Miocene and resulted in the overlapping of the Ligurian complex onto the Triassic evaporites, the rocks of the Verrucano group, and Paleozoic phyllites along the gently dipping normal faults. The third (latest) episode (Pliocene to the present) is marked by the development of the normal faults steeply dipping northeast. In this region, the surface heat flux reaches its maximal intensity (above 500 mW/m^2) along several normal faults (Bellani et al., 2004). This fact was interpreted as the result of the fluid motion in the localized zones of deformation corresponding to the main shear zone. This process is continuing up to the present and can be associated with both the liquid and overheated vaporous geothermal fluids.

The tectonic extensions in Tuscany were accompanied by acid magmatism. Drilling in the Larderello region exposed numerous acid dikes and granitoids with the age of cooling of 2.25–3.80 Ma, which thread through the Paleozoic mica schists and the gneiss complex. Intrusion of these magmatic bodies at some places resulted in overlapping of the contact-metamorphic mineral associations onto the earlier ones.

The deep structure of the Larderello region is traced by the seismic reflection data. A sharp contrast between the weakly reflecting upper crust and strongly reflecting middle and lower crust is observed along the seismic profiles. Based on the seismic data and the results of drilling, Brogi et al. (2003) supposed that the younger steeply dipping faults are the surface manifestations of the extensional shear deformations, which tend to flatten out with the increasing depth.

The deep structures in the Travale geothermal field are determined by the behavior of the so called H- and K-horizons located at the depths of 1–2 and 5–6 km, respectively (see their locations in Figure 6.2). Both horizons are marked by the high amplitudes and intermittent lateral distribution of seismic reflections (Bertini et al., 2005). Here, the H-horizon is more discontinuous than the K-horizon, which is located at a larger depth. The boreholes penetrating this horizon show that it is a fractured zone filled with vapor and located near the top portions of the granite intrusion.

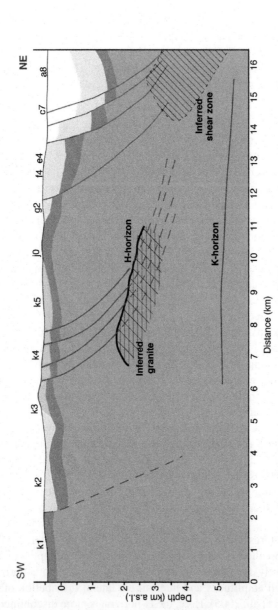

FIGURE 6.2 Geological section along A–A′ profile in Figure 6.1(b) (modified after Bellani et al., 2004). Colors define the same units of Figure 6.1. Purple color refers to the Larderello substratum composed by the Monticiano–Roccastrada unit and the Gneiss complex.

To all appearances, the intervals of reduced amplitudes against the background predominantly strong reflections are caused by the vapor-gaseous fluids permeating the fractured massifs (Capetti et al., 2005). The borehole measurements within this area support the correlation between the anomalous reflections and vapor production.

6.3 ELECTROMAGNETIC SOUNDING

6.3.1 MT Data

The magnetotelluric data in the Travale region were measured along the profile A–A' in the frequency band from 350 to 0.001 Hz at 69 observation sites, whose locations are shown in Figure 6.1b. The size of the studied area is about 4 × 4 km. The problem of noise was solved by conducting MT measurements at the remote base on the Sardinia Island. In order to remove static shift of the MT curves, the transient electromagnetic (TEM) sounding was carried out at eight MT sites. In addition to TEM data, the geological information retrieved from the outcrops of the rocks and the corresponding electrical resistivity estimates were used. Considering the fact that the average resistivity in the rock sequence is very low, the maximal sounding depth corresponding to the period of 1000 s is estimated at about 10 km.

6.3.2 Electrical Resistivity Cross-Section

By applying different algorithms, Manzella et al. (2006) carried out 1-D inversion of the MT data using the determinant of the impedance tensor as well as the joint 2-D inversion of the TE- and TM-modes with initial model based on the geological section along profile A–A' (Figure 6.2). The 2-D distribution of electrical resistivity along the A–A'profile (see Figure 6.1b for its location) yielded by the 2-D MT inversion is shown in Figure 6.3.

A steeply dipping conductive structure, which can probably be identified as a fault, is revealed in the southwest of the region. This structure stretches from the surface to the depth of the deep crustal conductor beneath the Travale area, whose roof coincides with the seismic K-horizon. The surface exposure of this steeply dipping structure corresponds to the Boccheggiano fault enriched with hydrothermal minerals. At the same time, by analyzing the resistivity data alone, one cannot find out whether this fault remains being an area of active fluid circulation (which has been certainly the case in the past), or it is a zone containing the accumulations of overheated steam.

The presence of two highly resistive areas, which are located at depths of 1.0–1.5 km and 3–5 km and are connected with each other, is the most prominent feature of the section shown in Figure 6.3. The increase in the electrical resistivity is a characteristic sign of the overheated vapor-gaseous mixture, whose chemical composition is presented by Valori et al. (1992): CO_2 (94.8 wt.%), CH_4 (1.2–3 wt.%), H_2S (1.6–2.9 wt.%), H_2 (0.7–2.2 wt.%), and

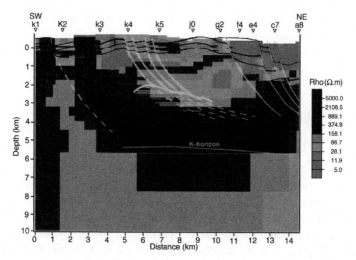

FIGURE 6.3   Electrical resistivity cross-section obtained by 2-D inversion of TE- and TM-mode data along profile A–A′ (after Manzella et al., 2006).

N_2 (0.8–1.3 wt.%). The mixture also contains He, Ar, O_2, and CO; however, the total concentration of these components is below 0.1 wt.%. In the central part of the section, the resistivity anomaly flattens out, following the faults revealed by seismic reflection data (Bertini et al., 2005).

6.4   TEMPERATURE MODEL

As it has been mentioned above, in order to more accurately interpret the resistivity section, it is necessary to build also the deep temperature model. To this end Spichak and Zakharova (2014) used the indirect EM geothermometer.

6.4.1   Temperature Data

A number of wells are drilled along the profile A–A′ in its closest vicinity and a few measurements are available from each well. Overall, these data comprise temperature records in the range of 53.8–437.5°C from a depth interval of up to 3775 m (Figure 6.4). By horizontal interpolation of these data along profile A–A′ Bellani et al. (2004) reconstructed the isotherms to a depth of slightly below 2 km (Figure 6.5). Dense concentration of the isotherms toward the surface is observed in the central part of this distribution. It is remarkable that their configuration coincides with the sloping subsurface high-resistivity anomaly in the resistivity cross-section shown in Figure 6.3.

6.4.2   Application of EM Geothermometer

The calibration of the EM geothermometer was carried out by setting the correspondence between the values of electrical resistivity and temperatures from 12

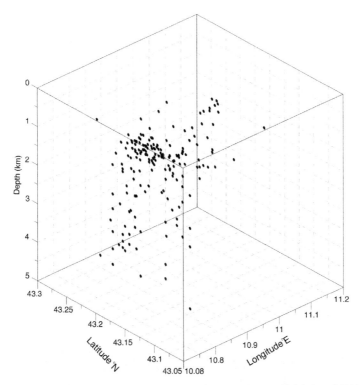

FIGURE 6.4 Locations of the temperature records in the study area (Spichak and Zakharova, 2014).

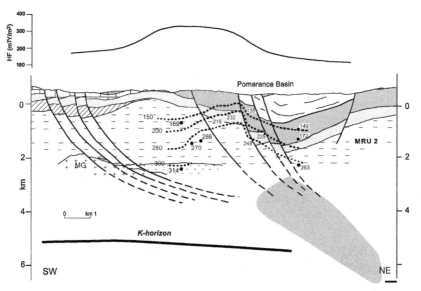

FIGURE 6.5 Geological cross-section, isotherms, and surface heat flow (upper graph) along the profile A–A′ (its location is given in Figure 6.1b): black circles denote the locations of the temperature measurements, isotherms (in °C) are indicated by dotted lines (after Bellani et al., 2004).

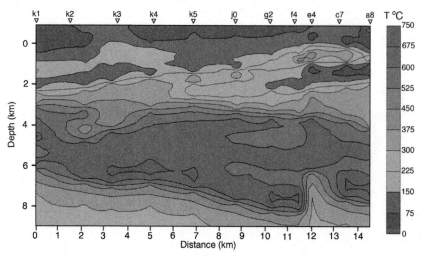

FIGURE 6.6   Temperature cross-section revealed from MT data along the profile A–A' (Spichak and Zakharova, 2014).

nearest boreholes (see Figure 6.1b for their locations). To this end the resistivity values were first estimated in the locations of the temperature records by means of ANN taught by the resistivity values from the 2-D model.

In order to assess the accuracy of the temperature forecasts by EM geothermometer Spichak and Zakharova (2014) have forecasted the temperature profiles in the locations of the geothermal exploration wells Travale_sud_1 (N1) and Montieri_4 (N4) situated at some distance from the profile A–A' using the resistivity profiles in the sites *j0* and *k4*, respectively (Figure 6.1b).

The results of the temperature forecast were compared with the temperature logs from the corresponding boreholes (Barelli et al., 2000). The average relative errors for the wells N1 and N4 were 15.0 and 11.3%, respectively. Taking into account quite large distances between the wells for which these temperature forecasts were carried out and locations of MT sites, from which the resistivity values were used for the temperature estimation, this accuracy could be considered as appropriate.

After calibration and testing, the EM geothermometer was applied for reconstructing the 2-D temperature model up to a depth of 10 km along the profile A–A' (Figure 6.6) using the resistivity values from the 2-D resistivity model (Manzella et al., 2006).

6.5   JOINT ANALYSIS OF THE RESISTIVITY AND TEMPERATURE MODELS

As it has been mentioned above, the electrical resistivity cross-section (Figure 6.3) was interpreted by Manzella et al. (2006) from the standpoint of searching for the zones of liquid fluids circulating along the faults in this region.

Correspondingly, the small conductive zone located in the central part of the section below the K-horizon was considered as the source for these fluids. However, the interpretation of resistivity cross-sections in geothermal terms always contains significant uncertainty. The point is that, depending on the predominant type of the fluid source (liquid or gas), these fluids cause low-resistivity or high-resistivity anomalies, respectively. This uncertainty can be reduced by a joint analysis of resistivity and temperature models.

The comparison of the obtained models of electrical resistivity and temperature (Figures 6.3 and 6.6, respectively) shows that the large highly resistive zone in the depth interval from 3.5 to 5.5 km, whose bottom coincides with the reflecting K-horizon, is located in the region of very high temperature exceeding 600°C at large depths. According to the seismic data (Fiordelisi et al., 2005), this anomaly can be associated with the presence of the overheated gas–vapor mixture (Fournier, 2007), which has high electrical resistivity.

Almost constant chemical composition of the fluid over a drilled area of approximately 400 km^2 (Giolito et al., 2007) and a relative gravimetric minimum (<20 mgal) over Travale area (Baldi et al., 1995) enables to interpret the large high-resistivity and high-temperature area observed from the depth of 3.5–5.5 km as a fractured granitic massif filled by supercritical mixture of gases and vapor. This supports the hypothesis of the presence at these depths of unconventional deep-seated geothermal resource (Bertini et al., 2005).

A remarkable fact is that a thin inclined highly resistive layer with temperature of 300°C stretches from this area to the surface (Figure 6.3). Until study carried out by Spichak and Zakharoba (2014), this structure was considered as an independent near-surface geothermal reservoir (see, for instance, Fiordelisi et al., 2005; Manzella et al., 2006). From the new standpoint, it can be interpreted as the pathway supplying hot vapor–gas mixture to the surface, which is supported by geochemical evidence (Giolito et al., 2007).

6.6 CONCLUSIONS

The joint analysis of the models of electrical resistivity and temperature up to a depth of 10 km suggests that the heat transfer in the Travale geothermal system is carried out by the overheated vapor–gas fluids rather than by liquid one, as it has been previously concluded from the model of the electrical resistivity only.

Another important result is that the structure that was previously identified as a pair of geothermal reservoirs (by the interpretation of the seismic and MT data) should probably be considered as a single reservoir having a subsurface branch.

Finally, it can be seen from the temperature model reconstructed using the EM geothermometer that the temperature below a depth of 3.5 km exceeds 500°C. Thus, it is reasonable to consider this high-resistivity and high-temperature reservoir as a potential unconventional deep-seated resource.

REFERENCES

Baldi, P., Bellani, S., Ceccarelli, A., Fiordelisi, A., Rocchi, G., Squarci, P., Taffi, L., 1995. Geothermal anomalies and structural features of southern Tuscany (Italy) (Expanded Abstr). World Geothermal Congress. Florence, Italy. pp. 1287–1291.

Barelli, A., Bertini, G., Buonasorte, G., Cappetti, G., Fiordelisi, A., 2000. Recent deep exploration results at the margins of the Larderello-Travale geothermal system (Expanded Abstr). World Geothermal Congress. Kyushu–Tohoku, Japan. pp. 965–970.

Bellani, S., Brogi, A., Lazzarotto, A., Liotta, D., Ranalli, G., 2004. Heat flow, deep temperatures and extensional structures in the Larderello geothermal field (Italy): constrains on geothermal fluid flow. J. Volc. Geotherm. Res. 132, 15–29.

Bertini, G., Casini, M., Ciulli, B., Ciuffi, S., Fiordelisi, A., 2005. Data revision and upgrading of the structural model of the Travale geothermal field (Italy) (Expanded Abstr). World Geothermal Congress. Antalya, Turkey.

Brogi, A., Lazzarotti, A., Liotta, D., Ranalli, G., 2003. Extensional shear zones as imaged by reflection seismic lines: the Larderello geothermal field (Central Italy). Tectonophysics 363, 127–139.

Capetti, G., Fiordelisi, A., Casini, M., Ciuffi, S., Mazzotti, A., 2005. A new deep exploration program and preliminary results of a 3D seismic survey in the Larderello-Travale geothermal field (Italy) (Expanded Abstr). World Geothermal Congress. Antalya, Turkey.

Fiordelisi, A., Moffatt, J., Ogliani, F., Casini, M., Ciuffi, S., Romi, A., 2005. Revised processing and interpretation of reflection seismic data in the Travale geothermal area (Italy) (Expanded Abstr). World Geothermal Congress. Antalia, Turkey.

Fournier, R., 2007. The physical and chemical nature of supercritical fluids (Expanded Abstr). ENGINE Workshop on Exploring High Temperature Reservoirs: New Challenges for Geothermal Energy, Volterra, Italy.

Giolito, C., Ruggieri, G., Gianelli, G., Manzella, A., 2007. The deep reservoir of the Travale geothermal area: mineralogical, geochemical and resistivity data (Expanded Abstr). ENGINE Workshop on Exploring high temperature reservoirs: new challenges for geothermal energy, Volterra, Italy.

Khutorskoi, M.D., Viskunova, K.G., Podgornykh, L.V., Suprunenko, O.I., Akhmedzyanov, V.R., 2008. A temperature model of the crust beneath the Barents Sea: investigations along geotraverses. Geotectonics 42, 125–136.

Limberger, J., Van Wees, J.D., 2013. European temperature models in the framework of GEOELEC: linking temperatures and heat flow data sets to lithosphere models (Expanded Abstr). European Geothermal Congress. Pisa, Italy.

Manzella, A., Spichak, V., Pushkarev, P., Kulikov, V., Oskooi, B., Ruggieri, G., Sizov, Yu., 2006. Deep fluid circulation in the Travale geothermal area and its relation with tectonic structure investigated by a magnetotelluric survey (Expanded Abstr). Workshop on Geothermal Reservoir Engineering. Stanford, USA.

Spichak, V.V., Zakharova, O.K., 2014. Gaseous vs aqueous fluids: Travale (Italy) case study using EM geothermometry (Expanded Abstr). XXXIX Workshop on Geothermal Reservoir Engineering, Stanford University, USA.

Valori, A., Cathelineau, M., Marignac, Ch., 1992. Early fluid migration on a deep part of the Larderello geothermal field: a fluid inclusion study of the granite sill from well Monteverdi 7. J. Volc. Geotherm. Res. 51, 115–131.

Chapter 7

Estimating Deep Heat Transfer Mechanisms: Soultz-sous-Forêts (France) Case Study

Chapter Outline

7.1 INTRODUCTION

Fluid circulation is associated with convective regime of the heat transfer, so, our ability to estimate somehow the dominant heat transfer mechanism (especially, at large depths not reachable by drilling) may help to locate the potential geothermal reservoir.

A number of geological, geophysical, and geochemical studies are provided in the Soultz-sous-Forêts geothermal zone (France) in the framework of the project on enhanced geothermal systems (EGS) (Genter et al., 2009). Temperature in this area is one of the most important physical parameters to estimate both at the exploration and exploitation stages. The use of an indirect electromagnetic geothermometer allows high accuracy temperature estimation at large depth based on the ground EM data and available temperature well logs. The studies carried out by Spichak et al. (2010, 2014) were aimed at estimating the dominant heat transfer mechanisms at large depths in the Soultz-sous-Forêts geothermal area and constraining the location for drilling new borehole. Below we discuss the obtained results following these papers.

Electromagnetic Geothermometry. http://dx.doi.org/10.1016/B978-0-12-802210-8.00007-1

7.2 GEOLOGICAL SETTING

The Soultz geothermal area is located within the Upper Rhine graben (Figure 7.1), which forms a part of the European Cenozoic rift system that extends in the foreland of the Alps from the Mediterranean to the North Sea coast. The Moho, which is the boundary between the earth's crust and the mantle, shows a topography of its depth thickness with a doming structure below the Upper Rhine graben.

Different geothermal wells are located inside the Upper Rhine graben and locally penetrate the so-called Soultz horst (Figure 7.1). (The location of the main geothermal borehole GPK2 considered in the following sections is marked in the upper panel.) It is made of sedimentary Cenozoic and Mesozoic formations (Triassic to Middle Jurassic) lying on a Paleozoic crystalline basement. At depth, two granite units are well-known: the porphyritic biotite-rich granite

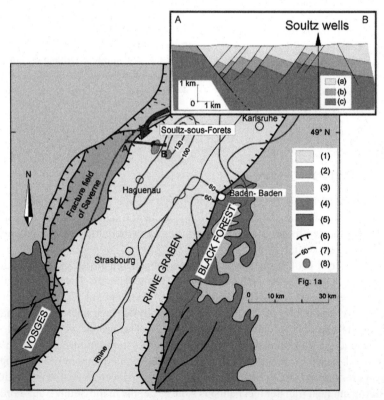

FIGURE 7.1 Location of the EGS Soultz site and geology of the Upper Rhine Graben: (1) Cenozoic sediments, (2) Jurassic, (3) Trias, (4) Permian, (5) Hercynian basement, (6) Border faults, (7) Temperature distribution in °C at 1500 m depth (Haenel et al., 1979), (8) Local thermal anomalies (Haenel et al., 1979). Simplified cross-section through the Soultz site: (a) Cenozoic filling sediments (b) Mesozoic sediments, (c) Paleozoic granite basement (after Dezayes et al., 2005).

and the fine-grained two-mica granite (Dezayes et al., 2005). Both sedimentary formations and Paleozoic granite are affected by a series of subvertical North–South normal faults dipping westward or eastward. Two major paleotectonic phases have been recognized in the Upper Rhine graben: an initial North–South compression (Eocene) is followed by an important East–West Oligocene extensional phase. The latter is mainly responsible for most observed structures and the actual geometry of faults and layers (Dezayes et al., 2005).

7.3 PREVIOUS TEMPERATURE ASSESSMENTS

During oil exploration, numerous temperature measurements have been made at depth in the Pechelbronn oil-bearing region (Haas and Hoffmann, 1929). Based on approximately 500 measurements, this old study shows that isotherms are influenced primarily by the tectonic structure of the Rhine graben. The hottest zone at 400 m depth is located along the western part of the Soultz horst and is characterized by NE–SW elongation (Haas and Hoffmann, 1929). It is remarkable that the configuration of the temperature contours mapped at a regional scale at the depth 1500 m (Figure 7.1 (7)) is similar to those mapped at local scale at the depth 400 m with the maximum located in the Soultz area, which seems to be attributed to hydrothermal fluid circulations (Haenel et al., 1979).

It is worth mentioning in this connection that the temperature contour maps are constructed at different depths mainly by linear interpolation of the temperature records available at that depth from a number of boreholes (Haenel et al., 1979; Schellschmidt and Clauser, 1996; Pribnow and Schellschmidt, 2000; Dezayes et al., 2005). Under the conditions of irregular distribution of the wells in the studied area and different depth ranges, where the temperature records are available (often very few), it is practically impossible to draw reliable vertical temperature cross-sections especially for the depths exceeding the depths of the drilled boreholes.

The only example known from the literature concerns the vertical temperature cross-section for the profile crossing the Upper Rhine graben area from NW to SE reconstructed in (Schellschmidt and Clauser, 1996; Pribnow and Hamza, 2000) by projecting of the 746 temperature values from 174 boreholes distributed in the 20 km vicinity from each side of it (Figure 7.2).

Despite of this cross-section gives a general idea about the vertical temperature distribution in the Upper Rhine graben area up to the depths 1000–1500 m, the accuracy of such temperature reconstruction for the Soultz area is doubtful, since it is implicitly based on the assumption on two-dimensionality of the temperature distribution in the SW–NE direction in the regional scale (at least, along 40 km). On contrary, the temperature map for the depth 800 m provided in (Pribnow and Schellschmidt, 2000) clearly indicates that the temperature contours do not manifest two-dimensionality 20 km apart the NW–SE profile crossing Soultz. Moreover, the maximal depth of the temperature contours is restricted by the lengths of the used boreholes (in the most cases bounded by

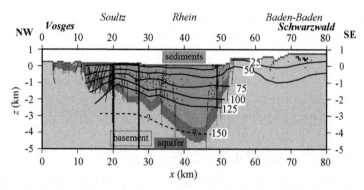

FIGURE 7.2 Model of the geological structures crossing the Rhine graben area along the profile A–B (Figure 7.1) based on seismic profiles together with temperature contours (in °C) from borehole measurements (after Pribnow and Hamza, 2000). Subvertical structures are fault zones. The projected temperature data positions are marked with dots (746 values from 174 boreholes). The 150°C is dashed because these high temperatures have only been measured at Soutlz ($X = 20$ km). Vertical black lines bound the study area.

1000–1500 m). So, in order to get more reliable temperature estimations for the Soultz geothermal area, in particular, for large depths, we have to consider alternative ways. Spichak et al. (2010, 2014) have used to this end an indirect EM geothermometer. Below we discuss this approach, estimate the accuracy of EM temperature extrapolation in depth and consider its application to the deep temperature assessment.

7.4   ELECTRICAL RESISTIVITY CROSS-SECTION

In order to get the resistivity profiles required for the indirect EM geothermometer calibration (i.e., for artificial neural network [ANN] training) the results of magnetotelluric (MT) survey carried out in 2007–2008 along the 13 km long W–E profile crossing the Soultz area (Geiermann and Schill, 2010) were used (appropriate MT sites are indicated in Figure 7.3).

The dimensionality analysis provided in (Geiermann and Schill, 2010; Schill et al., 2010) has shown that for the periods less than 1 s all MT data manifest 1-D dimensionality (the corresponding xy- and yx components practically coincide with each other – see two upper panels in Figure 7.4). For the periods less than 40 s the Swift's 2-D indicator (skew) was less than 0.2 (except the site nearest to the graben, where it was equal to 0.25 – see lower panel in Figure 7.4), indicating 2-D dimensionality and started to increase for longer periods giving the evidence of 3-D effects.

Based on the prior geological and geophysical information as well as on the results of forward MT modeling, Geiermann and Schill (Geiermann, 2009; Geiermann and Schill, 2010; Schill et al., 2010) have estimated the direction of two-dimensionality of the geological structure along N52°E (parallel to the graben border strike). 2-D inversion of both the TE- and the TM-mode of MT

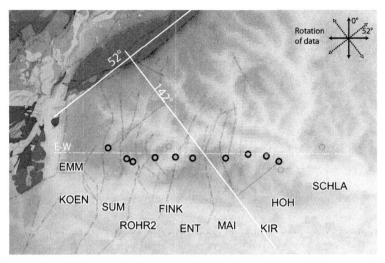

FIGURE 7.3   Map of the MT survey area (sites are marked by capital letters) (after Geiermann, 2009). White line labeled 52° approximates local graben strike, while line labeled 142° denotes the profile used for 2-D inversion. Thin lines denote projection of MT sites locations on that profile. Dashed white line indicates E–W profile. The Cross in the upper right corner depicts rotation of the data. The faults are plotted less opaque.

data for the periods less than 40 s was carried out using the code by Rodi and Mackie (2001) along the profile perpendicular to this direction (the distances between the MT sites being projected onto the new profile prior to the inversion). A prior model accounting for the geometry of the Rhine graben was used and its main elements were preserved during inversion. The error floor was 5% for both the resitivity and phase and the rms error was equal to 2.1, so, the average misfit of apparent resistivities was 10%. This result is a tradeoff between the model fit and a smoothness constraint (Tikhonov's regularization). For more details on the data processing, analysis and inversion we refer to Geiermann and Schill (2010), where these topics are broadly treated.

Figure 7.5 indicates the result of 2-D MT inversion carried out by Geiermann and Schill (2010). Generally, the Rhine graben model is well represented. A conductive zone between MT sites SUM and ROHR2 extends vertically over the Buntsandstein formation between Soultz and Kutzenhausen faults into the granitic basement (depths more than 2000 m). The "horst" part of the model is in a good agreement with the prior model accounting for the fault geometry of this area. The Triassic sediments coincide with an area of increasing resistivity from some tens to a few hundreds of Ωm.

The granitic basement is electrically more homogeneous and characterized by strong increase of the resistivity. In accordance with results of the residual Bouguer anomaly inversion (Geiermann and Schill (2010), it is marked by a comparatively low mean density of about 2500 kg/m^3 possibly indicating a strong fracturing at large depths.

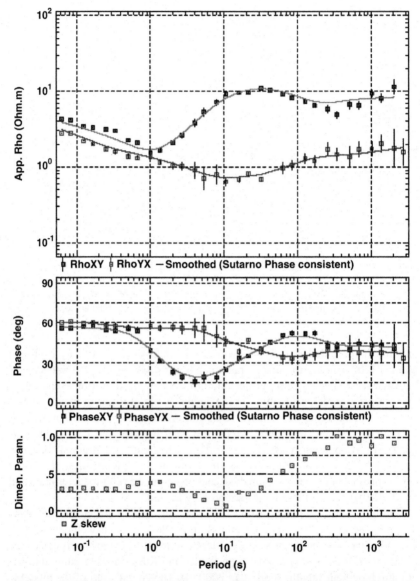

FIGURE 7.4   MT data for the site EMM nearest to the graben shoulder (see Figure 7.3 for its location) (redrawn from Geiermann (2009)).

7.5   GEOTHERMOMETER VALIDATION

Before building of the temperature model based on MT data the EM geother-mometer was tested in three ways. First, the forecasted temperature profiles were compared with temperatures logs in the framework of so-called "retro-modeling"; second, the dependence of the temperature estimation accuracy on

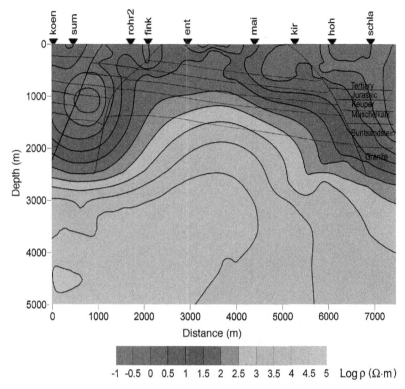

FIGURE 7.5   2-D resistivity cross-section along the NW–SE profile perpendicular to the 2-D dimensionality axis revealed by inversion of rotated MT data (after Spichak et al., 2014). The fault geometry and formation boundaries are redrawn from (Schill et al., 2009).

the ratio between the borehole length and extrapolation depth was studied; and, finally, the affect of the resistivity uncertainty on the temperature assessment accuracy was estimated.

7.5.1   Retro-Modeling

Validation of the temperature assessment using MT data was fulfilled, first, by retro-modeling. The point is that drilling of the borehole GPK2 (see Figure 7.6 for its location) was carried out in three stages: after an early phase of exploration by drilling up to the shallow depth (2000 m) it was deepened, first, up to the depth of 3878 m and later additionally up to the depth of 5046 m (Genter et al., 2009) (Figure 7.7). Accordingly, we tested the EM geothermometer by means of comparison of the forecasted temperature profile with real temperature log as if it was done before appropriate drilling from the depths of 2000 and 3878 m.

Each time EM geothermometer was calibrated by the correspondence between appropriate parts of the temperature log, from one hand, and modeled

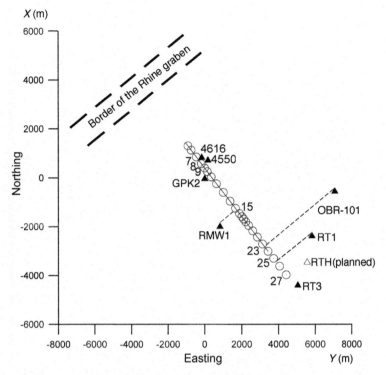

FIGURE 7.6 Schematic map of the survey area (after Spichak et al., 2010). Circles are the grid nodes of the 2-D mesh along the rotated profile; black triangles indicate the adjacent boreholes the logs from which were used for the indirect EM geothermometer calibration.

resistivity profile nearest to the borehole GPK2 (marked as "9" in Figure 7.6), on the other hand. After calibration the geothermometer was applied for successive temperature extrapolation downward using the modeled resistivity values from the deeper part of the profile (Figure 7.7).

Figure 7.7 shows the results of the EM temperature forecast "after drilling" of 2000 and 3878 m in comparison with the temperature well log. The relative forecast error for the extrapolation from the upper 2 km to the depth range up to 3878 m was only 1.8%, while in the latter case it was equal to 0.4%. It is worth mentioning in this connection that high extrapolation accuracy is caused by favorable ratios between the lengths of the temperature profiles used for calibration, on the one hand, and the extrapolation depth ranges, on the other hand (see Section 4.3). In the next section we study this effect in more details.

7.5.2 Accuracy Dependence on Extrapolation Depth

Spichak and Zakharova (2009) have shown that under the condition that the EM soundings are carried out close to the wells from which the temperature

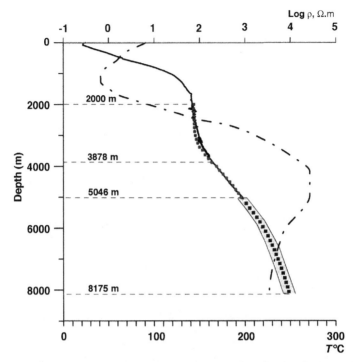

FIGURE 7.7 Estimated temperature profiles (dotted line) in the well GPK2 for the depth ranges 2000–3878, 3878–5046, and 5046–8175 m (after Spichak et al., 2010, 2014). The temperature well log is indicated by solid line, profile of the resistivity logarithm is marked by dashed–dotted line. Hatched area indicates the bars corresponding to 10% uncertainty in the resistivity model used for the temperature forecast.

data are taken for calibration of the geothermometer, the relative temperature extrapolation accuracy does not explicitly depend on any external factors. Actually, it depends only on the ratio between the extrapolation depth and length of the borehole.

The accuracy of the temperature extrapolation in depth was estimated using the temperature records measured in the boreholes (see Table 7.1 for their characteristics) and adjacent resistivity profiles from the 2-D resistivity model. The circles in the Figure 7.6 indicate locations of the grid nodes at the surface, the numbers nearby (7, 8,…,27) mark the selected resistivity profiles actually used for the geothermometer calibration, while the black triangles denote the related boreholes, the temperature logs from which were used to this end.

In the first experiment six artificial neuronets were trained in correspondence between the resistivity logarithm profiles and temperature well logs (upper parts of the corresponding graphs in Figure 7.8). The whole length of each borehole was divided into 10 intervals and ANNs were taught in the

TABLE 7.1 Characteristics of the temperature data used for calibration

Borehole	Type of records	Depth range (m)	Number of records	Nearest resistivity profile
4616	Log	1–1409	–	7
4550	Log	1–1494	–	8
GPK2	Log	1–5046	–	9
RMW1	Local temp.	200–709	9	15
OBR101	Local temp.	300–1778	15	23
RT1	Local temp.	545–1850	10	25
RT3	Local temp.	558–1540	14	27

After Spichak and Zakharova (2014).

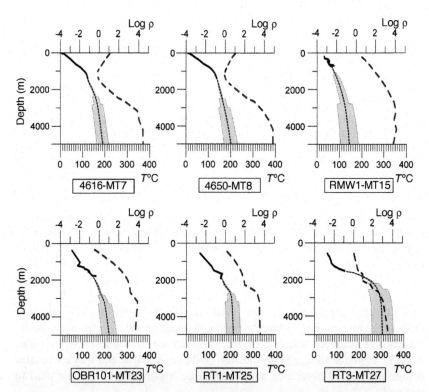

FIGURE 7.8 Profiles of the resistivity logarithm from the 2-D model (dashed lines) and the temperature profiles in the adjacent boreholes (solid line – well log, dotted line – extrapolated temperature profile, hatched areas indicate the bars, corresponding to 10% uncertainty in the resistivity model used for the temperature forecasting) (after Spichak et al., 2010, 2014).

TABLE 7.2 Relative errors (%) of the temperature extrapolation in depth depending on the portion δ of the temperature well logs and nearest resistivity profiles used for the calibration

δ	4616-MT7	4550-MT8	RMW1-MT15	OBR101-MT23	RT1-MT25	RT3-MT27
0.1	78.7	77.8	43.0	52.7	45.7	39.7
0.2	69.1	54.9	36.5	51.7	43.7	32.3
0.3	10.1	41.2	11.6	27	38.8	27.9
0.4	5.2	21.6	8.5	15.3	30.2	23.7
0.5	0.9	7.4	6.4	13.5	11.8	18.1
0.6	1.5	6.2	4.8	14.8	10.7	21.7
0.7	0.8	4.1	5.8	12.7	13.2	21.4
0.8	0.6	1.7	6.6	9.5	13.2	18.8
0.9	0.7	1.7	5.9	4.2	12.6	10.5

After Spichak and Zakharova (2014).

correspondence between the resistivity logarithm and temperature values determined within the upper δ (1/10, 2/10, ... 9/10) parts of the profiles for each well. After each training, the neuronets were used for the temperature forecasting to the remaining parts of the profiles and comparing the results with actual temperature values. Accordingly, the input vector of ANN consisted from the depths and resistivities taken from the δ-th part of the corresponding profiles, while the output vector consisted from the appropriate temperatures recorded at the same depths.

Shown in Table 7.2 are the errors of the temperature estimation at different depths depending on the portion δ of the temperature and logarithm resistivity profiles (from the top to maximal well lengths) used for calibration. It can be seen from Table 7.2 that for all wells the testing errors ε decrease with increasing δ (on average, from ~56% at $\delta = 0.1$–6% at $\delta = 0.9$).

In Figure 7.9, a plot is shown illustrating the dependence of the mean relative error ε of the neuronet extrapolation of temperature in depth (based on MT data measured at the site closest to the well versus the portion of profiles used for neuronet training). From this plot one can conclude that extrapolation to the depths, say, twice as large as the borehole lengths results in errors less than 10%. Assuming that the accuracy estimates obtained in testing of the geothermometer in the depth range 0–2 km remain valid also in a bigger depth scale (Spichak and Zakharova, 2009), we could assess the potential relative temperature uncertainties when consider the depth range 0–5 km beneath other wells using the graph and appropriate bars from Figure 7.9.

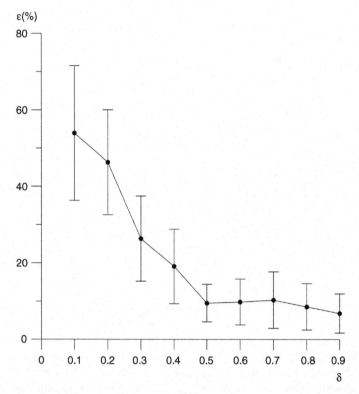

FIGURE 7.9   Dependence of the average relative error ε of the EM temperature extrapolation on the portion δ of the temperature well logs and the appropriate resistivity profiles used for the neuronet training (after Spichak et al., 2010).

7.5.3   Effect of the Resistivity's Uncertainty

The accuracy of the MT data inversion evidently affects the accuracy of the temperature estimation based on the derived resistivity values. Indeed, due to nonuniqueness of the inverse geophysical problems the resulting resistivity distribution (even if it fits the data) could be quite ambiguous and requires some regularization using external constrains. Moreover, it could be affected also by incorrect hypothesis on the dimensionality of the geological medium (say, 1-D or 2-D instead of 3-D), features of the algorithm and appropriate software used for the inversion, application of starting models being far from reality, and so on. The factors enumerated above may finally disturb the resistivity model in unpredictable way, which, in turn, could affect the temperature forecast based on the indirect EM geothermometer.

In order to model the affect of the resistivity's uncertainty on the temperature accuracy a technique rather common in computational physics was used. The resistivity profile 9 from the 2-D resistivity model (indicated as 9 in Figure 7.6) was successively altered by adding of a synthetic Gaussian noise with a zero mean value and standard deviations $\sigma = 5, 10,\ldots, 50\%$. Each disturbed profile was used then for the temperature extrapolation in the adjacent borehole GPK2 (having the longest temperature well log ranging from 0 to 5046 m). By analogy with the previous section the ANNs were taught by the distorted resistivity values and corresponding temperature data from the well log so that each time only a part ($\delta = 0.1, 0.2,\ldots, 0.9$) of the profiles was used for training and the remaining part was used for testing by comparison of the forecasted temperatures with real ones.

Table 7.3 shows the resulting errors of the temperature forecast for each value of the noise level (σ) in the resistivity data depending on the portion of the profiles (δ) used for the thermometer calibration. The comparative study indicates that they depend both on the portion of the profiles (δ) used for calibration and on the resistivity's uncertainty (σ). However, while the former dependence is similar to that obtained in the previous studies (see, in particular, the previous section), the latter one is at the first glance rather surprising: for $\delta \geq 0.5$ (i.e., when the extrapolation depth is less than twice bigger than the borehole length) the temperature forecasting errors are monotonically increasing with σ (though remain less than 20% even for $\sigma = 50\%$), while for $\delta < 0.5$ (i.e., when the extrapolation depth is more than twice bigger than the borehole length) the temperature forecasting errors practically do not depend on σ and remain at the level less than 20% (except the value $\delta = 0.1$, when they exceed 30%, if σ is bigger than 30%).

The latter results could be explained as follows. The ANN recognition is based on the similitude principle, that is, in contrary to common inversion procedures minimizing the affect of errors in the testing data, the best ANN recognition results are achieved when the training and testing data are similar in general sense. This, in particular, means that minimal errors of the temperature forecast in depth correspond to situations when both the training and the testing resistivity data are equally disturbed (the level of noise does not matter) and this is our case. It is interesting to note in this connection that Spichak (2007) has suggested adding of the artificial noise to the training resistivity data in order to minimize the errors of the ANN recognition from noisy EM data. The errors of the ANN forecast were minimal, when the training data were disturbed by the artificial noise with the same standard deviation as it was observed in the testing data.

We can infer, first, that while modeling of the resistivity's uncertainty by adding of the synthetic Gaussian noise to the resistivity model may not provide exact values of the temperature errors corresponding to different types of the resistivity's ambiguity enumerated above, it still gives a general idea about their behavior depending on the level of uncertainty and the extrapolation depth. Second, since the ANN temperature forecasting is based on the similitude principle, the same resistivity's uncertainty both in the training and testing resistivity data very weakly affects the resulting temperature, the latter being influenced mainly

TABLE 7.3 Relative errors of the temperature extrapolation in depth depending on standard deviation (σ) of the Gaussian noise added to the resistivity data and the portion (δ) of the temperature well logs and nearest electrical resistivity profiles used for calibration

	Depth range, m		σ, %										
δ	Calibration	Forecast	0	5	10	15	20	25	30	35	40	45	50
0.1	0–500	500–5000	22.7	23.8	28.5	28.8	29.1	28.9	27.7	31.9	34.5	37.6	41.4
0.2	0–1000	1000–5000	17.4	18.2	15.2	14.9	15.1	14.4	15.3	15.1	14.8	14.9	15.2
0.3	0–1500	1500–5000	16.4	15.4	16.2	16.3	16.4	16.5	16.7	16.8	16.8	16.8	16.9
0.4	0–2000	2000–5000	15.1	15.3	15.9	15.9	16.0	15.6	16.4	15.7	15.7	15.7	15.7
0.5	0–2500	2500–5000	4.3	7.8	9.3	11.6	14.7	16.6	17.4	17.5	18.0	18.4	19.3
0.6	0–3000	3000–5000	3.8	2.0	2.7	4.7	8.4	10.1	14.2	15.6	16.1	16.5	16.5
0.7	0–3500	3500–5000	0.9	1.1	3.0	3.6	6.0	8.1	10.3	13.3	13.7	14.0	13.7
0.8	0–4000	4000–5000	1.0	1.2	2.2	1.4	0.7	2.0	3.0	4.2	4.5	4.7	5.2
0.9	0–4500	4500–5000	0.4	0.5	0.5	1.5	1.3	0.8	0.3	0.4	0.7	0.7	0.5

After Spichak and Zakharova (2014).

by the extrapolation depth (especially, when it exceeds the borehole length more than twice).

7.6 TEMPERATURE CROSS-SECTION

Temperature maps built earlier for the depths 400 and 1500 m manifest two-dimensionality of the temperature contours (at least, in the local scale) approximately in the same (SW–NE) direction (see Figure 7.1 for the temperature isolines at the depth 1500 m). This justifies indirect temperature estimations in the vicinity of the rotated survey line using vertical resistivity profiles from the model determined in the nodes of 2-D grid as a result of MT data inversion.

Construction of the vertical temperature cross-section along the NW–SE profile up to the depth 5000 m not reachable by the wells was made in four stages. First, temperature logs in the boreholes 4616, 4550, RMW1, OBR101, RT1, and RT3 located in the vicinity 2–4 km from the rotated profile (see Figure 7.6 for their locations and Table 7.1 for appropriate characteristics) were extrapolated. To this end the nearest vertical resistivity profiles from 2-D model crossing the horizontal plane at the locations indicated by circles (Figure 7.6) were used.

After training of six ANNs in correspondence between the resistivity logarithm and temperature values at the same depths, they were applied for the temperature extrapolation up to the depth of 5 km. Figure 7.8 shows the extrapolated temperature profiles (dotted lines) with appropriate error bars (hatched areas). They were determined according to Tables 7.2 and 7.3 and assuming 10% resistivity's uncertainty (see section 7.5.3 above).

It is seen from Figure 7.8 that maximal temperature uncertainty at the depth of 5 km is less than ~30 °C (beneath boreholes 4616, 4550, OBR101, and RT1), while it exceeds this value beneath the boreholes RMW1 (~40°C) and RT3 (~50°C), which are characterized by less favorable ratios between the borehole lengths and the extrapolation depths.

At the second stage another ANN was created in order to interpolate the resistivity values determined in the nodes of the 2-D grid into the locations of the temperature records in the extrapolated logs (see above) plus GPK2 log. Third, another ANN was taught by the correspondence between the resistivity determined in the temperature locations at the previous stage, coordinates and the appropriate temperature values. Finally, the vertical temperature cross-section in the study area (Figure 7.10) was built by means of the ANN created at the latter stage. The input of this ANN consisted of the resistivity values determined in the nodes of 2-D grid and their coordinates.

As it is seen from Figure 7.10, the resulting temperature cross-section essentially differs from that reconstructed by Pribnow and Schellschmidt (2000) and by Pribnow and Hamza (2000) (see Figure 7.2) based on averaging the temperature records from all available boreholes at the distance 20 km from each side of the NW–SE profile. On contrary to the latter one, it is more heterogeneous and clearly indicates two temperature anomalies located in the granitic basement

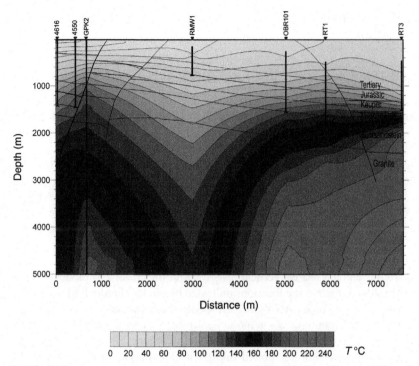

FIGURE 7.10   Temperature cross-section along the rotated NW–SE MT profile revealed from the resistivity data (after Spichak et al., 2010, 2014). Triangles indicate projections of the boreholes' locations onto this profile. The fault geometry and formation boundaries are redrawn from (Schill et al. 2009).

beneath boreholes GPK2 and RT1/RT3 the latter being more shallow. They could be interpreted as being caused by two big fluid circulating loops with a temperature minimum between them beneath the borehole RMW1, where the extrapolated temperature at the depth 5 km ranges between 105 and 185°C (Figure 7.8). In the two adjacent boreholes (GPK2 and OBR101) the temperature ranges at this depth are 195–205°C and 190–255°C (Figures 7.7 and 7.8, accordingly). Since both adjacent minima are higher than the maximum beneath RMW1, it could be assumed that the temperature minimum beneath this borehole is not caused by lack of temperature information at the depths larger than the borehole length.

Another remarkable feature of the temperature cross-section concerns to the isotherms' sinusoidal shape in the horizontal direction that supports the hypothesis on the deep-rooted fluid circulation in the fractured granitic basement (Clauser and Villinger, 1990; Le Carlier et al., 1994) along the faults' network, which could serve for the upward heat transfer. This finding agrees with conclusion of Schellschmidt and Clauser (1996) regarding "free convection within a single fault plane as driving mechanisms for an additional advection of heat" and with the concept of using the temperature as a tracer for fluid flow formulated by Pribnow and Schellschmidt (2000).

7.7 DOMINANT THERMAL REGIME AT LARGE DEPTH

In order to estimate the dominant thermal regime at a depth below the deepest borehole (GPK2), the indirect EM geothermometer was calibrated using the resistivity and temperature data from this borehole for the depth range 0–5046 m and used then for the temperature extrapolation to the depth range 5046–8175 m. Figure 7.7 shows the temperature well log and the resistivity logarithm profile 9 from 2-D resistivity model (see Figure 7.3 for location of its projection onto the horizontal plane). The extrapolated temperature is shown with error bars determined from the analysis carried out above under the assumption that the resistivity uncertainty in 2-D model corresponds to the standard deviation of 10%.

The analysis of the extrapolated profile between the depths 5000 and 6000 m indicates that the temperature gradient continues to be as high as in the depth range 3700–5100 m. This agrees with the temperature forecast for this depth range made by Pribnow et al. (1997) using linear extrapolation of the maximal heat flow gradient ($q = 59$ mWm^{-2}) observed at depths 3.5–3.7 km under assumption on conductive heat transfer mechanism. In particular, according to these authors, the temperature at the depth 6000 m could reach 217.3°C (appropriate error bars are not available), while according to EM geothermometry it ranges in 215–227°C. It is worth mentioning in this connection that the latter result is obtained without prior assumption on the conductive heat transfer mechanism at these depths.

According to EM temperature extrapolation, the temperature gradient slowly decreases down from the depth of 6000 m (Figure 7.7). This may indicate that the thermal regime switches again from conductive (at the depth range 3700–6000 m) to convective one. This convex curve may indicate the occurrence of a convective cell at a large depth in the granitic basement where according to residual Bouguer anomaly inversion (Geiermann and Schill, 2010) a strong fracturing may occur.

Based on chemical and isotopic fluid composition studies collected in various Soultz wells, different authors (Pauwels et al., 1993; Aquilina et al., 1997; Sanjuan et al., 2010) are favorably disposed toward the existence of a deep and hot geothermal reservoir with temperatures in the range 220–260°C. The concordant temperature values recorded by chemical geothermometers, which are slightly higher than the bottom-hole temperature measured in the well GPK2 at 5 km depth (close to 200°C), indicate the existence of chemical equilibrium reactions between the deep native reservoir brine and a mineralogical assemblage at 220–240°C (Sanjuan et al., 2010).

However, the hypothesis on the sedimentary origin of the geothermal fluids provided by these authors is not in agreement with the thermal estimates of our study. Indeed, the deepest part of the Soultz site (>5000 m) is thought to be within crystalline fractured basement and not within deep sedimentary Triassic

Buntsandstein formations (Figure 7.1, inset panel) as proposed by Sanjuan et al. (2010). Those sedimentary reservoirs are probably located in the central part of the Rhine graben. The Li isotopic composition of the native brine is not characteristic of a granite but would rather fit the sedimentary formations made of carbonate (Sanjuan et al., 2010), which could also be found in the central part of the Rhine graben (Jurassic, Muschelkalk).

Geochemical modeling of the fluids showed that the geothermal brine is close to equilibrium at 230–240°C with respect to quartz, albite, K-feldspar, calcite, dolomite, $CaSO_4:0.5H_2O$, fluorite, muscovite (illite), and smectites. Most of these minerals have already been observed in a hydrothermal alteration sequence in the Soultz granite (Ledesert et al., 1999). The occurrence of organic compounds in the deep native fluids and in the crystalline fracture zone (Aquilina et al., 1997; Ledesert et al., 1996, 1999) is an additional argument, which supports the hypothesis on a large-scale circulation between sediments and deep basement fractured rocks.

7.8 CONSTRAINING LOCATION FOR NEW BOREHOLE DRILLING

The selection of appropriate locations for geothermal drilling is one of the most important problems of the geothermal exploration. It is usually made under conditions of uncertainty regarding the temperature and permeability distribution in the studied area. Meanwhile, the range of the temperature uncertainty could be reduced by means of using the indirect temperature assessment from the EM sounding data.

At Rittershoffen (RTH), where a new well was planned to be drilled (see its preliminary location in Figure 7.6), Spichak et al. (2010, 2014) have compared the temperature profiles obtained by different ways:

- since RTH is situated just between the boreholes RT1 and RT3, its temperature profile could be close to either of their profiles extrapolated above using EM thermometry;
- under the assumption that the temperature varies linearly in the space between these profiles, the temperature profile in a RTH site could be determined by averaging of two abovementioned profiles in the boreholes RT1 and RT3 (Figure 7.11, line with black dots).

As it is seen from Figure 7.11, the profiles at the sites RT1 and RT3 differ from each other both in the shallow and deeper (extrapolated) parts the main difference being the negative temperature gradient revealed in the site RT3 at the depth range 1800–2500 m (though it switches again to the positive one at the depths deeper than 2800 m (compare with Figure 7.8). Comparison of three profiles leads to conclusion that in order to reach the temperatures around, say, 180°C, at the minimal depth the best plan to be followed would be to deepen the existing borehole RT3 or to drill a new one in its close vicinity.

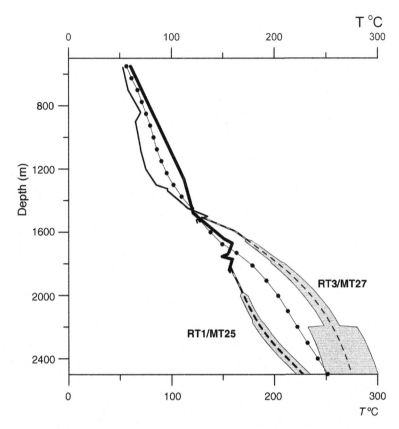

FIGURE 7.11 The indirect temperature forecast for the new borehole planned to be drilled in the RTH (after Spichak et al., 2010, 2014). The temperature in the borehole RT1 is marked by thick line, in RT3 – by thin line; temperatures extrapolated in these boreholes using nearest resistivity profiles are marked by dashed lines; the temperature profile averaged from the RT1 and RT3 well logs and their extrapolations – by line with dots. Hatched areas indicate the bars corresponding to 10% uncertainty in the resistivity model used for the temperature forecast.

7.9 CONCLUSIONS

It is shown that the indirect EM temperature estimation in the geothermal area with highly conductive sediments could be carried out using the resistivity cross-section resulting from 2-D MT data inversion. Validation of the temperature assessment fulfilled by comparison of the forecasted temperature profile with temperature log from the deepest borehole has resulted in the relative extrapolation accuracy less than 2%. It is found that the resistivity's uncertainty caused by MT inversion errors and by possible effects of external factors very weakly affects the resulting temperature the latter being influenced mainly by ratio between the borehole length and the extrapolation depth.

On contrary to the temperature model, reported for the Soultz area earlier, the vertical temperature cross-section reconstructed up to the depth of 5 km using the indirect EM geothermometer is more heterogeneous and clearly indicates two temperature anomalies located at depths beneath boreholes GPK2 and RT1/RT3 the latter being more shallow. The temperature contour map manifests sinusoidal behavior at the depths below 1.5 km, which agrees with the hypothesis on the convective heat transfer mechanism dominating at these depths.

The temperature profile estimated below the deepest borehole GPK2 indicates that the temperature gradient being linear in the depth range 3700–6000 m becomes slowly decreasing down from the depth 6000 m. This, in turn, may indicate that the thermal regime switches again from the conductive to convective one, which may be caused by deep-rooted fluid circulation in the fractured granitic basement and the heat or mass transfer along adjacent faults.

Finally, it is shown that the indirect EM geothermometer could be used for constraining the locations for drilling of new boreholes for geothermal exploration. In particular, its application to forecasting the temperature profiles in the RTH site offers estimates which constrain the site location for future drilling.

REFERENCES

Aquilina, L., Pauwels, H., Genter, A., Fouillac, C., 1997. Water-rock interaction processes in the Triassic sandstone and the granitic basement of the Rhine Graben: Geochemical investigation of a geothermal reservoir. Geochim. Cosmochim. Acta 61, 4281–4295.

Clauser, C., Villinger, H., 1990. Analysis of conductive and convective heat transfer in a sedimentary basin, demonstrated for the Rheingraben. Geophys. J. Int. 100, 393–414.

Dezayes, C., Genter, A., Hooijkaas, G., 2005. Deep-seated geology and fracture system of the EGS Soultz reservoir (France) based on recent 5km depth boreholes (Expanded Abstr). World Geothermal Congress. Antalya, Turkey.

Geiermann, J., 2009. 2-D magnetotelluric sounding and modeling at the geothermal site Soultz-sous-Forêts. Dipl. Phys., J. Gutenberg Universitat, Mainz, Germany, 98 pp.

Geiermann, J., Schill, E., 2010. 2-D Magnetotellurics at the geothermal site at Soultz-sous-Forêts. Comptes Rendus Geoscience 342 (7–8), 587–599.

Genter, A., Fritsch, D., Cuenot, N., Baumgartner, J., Graff, J.-J., 2009. Overview of the current activities of the European EGS Soultz project: from exploration to electricity production. Expanded Abstr. XXXIV Workshop on Geothermal Reservoir Engineering, Stanford University. Stanford, USA.

Haas, J.-O., Hoffmann, C.R., 1929. Temperature gradient in Pechelbronn oil bearing region, Lower Alsace: its determination and relation to oil reserves. Bull. Amer. Assoc. Petr. Geol XIII (10), 1257–1273.

Haenel, R., Legrand, R., Balling, N., Saxov, S., Bram, K., Gable, R., Meunier, J., Fanelli, M., Rossi, A., Salmone, M., Taffi, L., Prins, S., Burley, A.J., Edmunds, W.M., Oxburgh, E.R., Richardson, S.W., Wheildon, J., 1979. Atlas of subsurface temperatures in the European Community. Th. Schafer Druckerei GmbH, Hannover, Germany.

Le Carlier, C., Royer, J.-J., Flores, E.L., 1994. Convective heat transfer at Soultz-sous-Forêts geothermal site: implications for oil potential. First Break 12 (11), 553–560.

Ledesert, B., Berger, G., Meunier, A., Genter, A., Bouchet, A., 1999. Diagenetic-type reactions related to hydrothermal alteration in the Soultz-sous-Forêts granite. France Eur. J. Miner. 11, 731–741.

Ledesert, B., Joffre, J., Amblès, A., Sardini, P., Genter, A., Meunier, A., 1996. Organic matter in the Soultz HDR granitic thermal exchanger (France): natural tracer of fluid circulations between the basement and its sedimentary cover. J. Volcanol. Geotherm. Res. 70, 235–253.

Pauwels, H., Fouillac, C., Fouillac, A.-M., 1993. Chemistry and isotopes of deep geothermal saline fluids in the Upper Rhine graben: origin of compounds and water-rock interaction. Geoch. Cosmoch. Acta 57, 2737–2749.

Pribnow, D., Engelking, U., Schellschmidt, R., 1997. Temperature prediction for the HDR Project at Soutz-sous-Forêts. GGA tech. rpt. 115869, Hannover, 10 pp.

Pribnow, D., Hamza, V., 2000. Enhanced geothermal systems: new perspectives for large scale exploitation of geothermal energy resources in South America (Expanded Abstr). XXXI International Geological Congress. Rio-de-Janeiro, Brasil.

Pribnow, D., Schellschmidt, R., 2000. Thermal tracking of upper crustal fluid flow in the Rhine graben. Geophys. Res. Lett. 27 (13), 1957–1960.

Rodi, W., Mackie, R., 2001. Nonlinear conjugate gradients algorithm for 2D magnetotelluric inversion. Geophysics 66, 174–187.

Sanjuan, B., Millot, R., Dezayes, C., Brach, M., 2010. Main characteristics of the deep geothermal brine (5km) at Soultz-sous-Forêts (France) determined using geochemical and tracer test data. Geoscience 342, 546–559.

Schellschmidt, R., Clauser, C., 1996. The thermal regime of the Upper Rhine graben and the anomaly at Soultz. Z. Angew. Geol 42, 40–44.

Schill, E., Geiermann, J., Kümmritz, J., 2010. 2-D magnetotellurics and gravity at the geothermal site at Soultz-sous-Forêts (Expanded Abstr). World Geothermal Congress. Bali, Indonesia .

Schill, E., Kohl, T., Baujard, C., Wellmann, J.F., 2009. Geothermische Ressourcen in Rheinland-Pfalz: Bereiche Süd- und Vorderpfalz, 55. Ministerium für Umwelt Forsten und Verbrauch-erschutz, Mainz.

Spichak, V.V., 2007. Neural network reconstraction of macro-parameters of 3-D geoelectric structures. In: Spichak, V.V., ed. Electromagnetic sounding of the earth's interior. Elsevier, pp. 223–260.

Spichak, V.V., Geiermann, J., Zakharova, O., Calcagno, P., Genter, A., Schill, E., 2010. Deep temperature extrapolation in the Soultz-sous-Forêts geothermal area using magnetotelluric data. Expanded Abstr. XXXV Workshop on Geothermal Reservoir Engineering. Stanford University, USA.

Spichak, V.V., Geiermann, J., Zakharova, O., Calcagno, P., Genter, A., Schill, E., 2014. Estimating deep temperatures in the Soultz-sous-Forêts geothermal area (France) from magnetotelluric data. Near Surf. Geophys. (in press, ID nsg-2014-1132).

Spichak, V.V., Zakharova, O.K., 2009. The application of an indirect electromagnetic geothermometer to temperature extrapolation in depth. Geophys. Prosp. 57, 653–664.

Chapter 8

A New Conceptual Model of the Icelandic Crust: Hengill Case Study

8.1 INTRODUCTION

There are two main hypotheses on the structure of the unique Icelandic crust. The "thin and hot" model of the crust (Björnsson et al., 2005; Björnsson, 2008) is based on the implicit assumptions on the conductive heat flow and linear increase of the borehole temperature gradient in the lithologically uniform crust (Hermance and Grillot, 1974; Flóvenz, 1985; Flóvenz and Saemundsson, 1993; Tryggvason et al., 2002). According to it the maximal depth (5–6 km) of the most seismic events corresponds to the brittle/ductile transition for the basalt ($T = 650°C$), that is, at the upper/lower crust boundary, while the lower boundary of the high electrical conductivity layers (10–15 km) detected by magnetotelluric soundings is attributed to the base of the crust.

The alternative hypothesis (so-called "thick and cold" crustal model) is based mainly on the observation of a gradual increase of seismic P-wave velocities with depth and a stepwise rise from 7.2 to 7.7 km/s at about 22 km, which supposedly marks the Moho (Pavlenkova and Zverev, 1981; Bjarnason et al., 1993; Menke and Levin, 1994; Menke and Sparks, 1995; Menke et al., 1995, 1996; Foulger et al., 2003).

Electromagnetic Geothermometry. http://dx.doi.org/10.1016/B978-0-12-802210-8.00008-3
149

Unfortunately, these models offer no clue as to the following questions:

- what is the nature of highly conductive layers recognized by MT sounding at the depths of 1–3 and 10–15 km (Björnsson et al., 2005)?
- why has the drilling in the Krafla geothermal field penetrated rhyolitic magmas (~74% SiO_2, 1–2% H_2O) with a temperature of $T = 1100°C$ at a depth of 2.1 km and with a temperature $T = 386°C$ at a depth of 2.6 km (Elders and Fridleifsson, 2010)?
- why is the continuous seismic activity in this region mainly confined to the superposition of the zone constrained between the meridians 21.31° and 21.33°W and a band running beneath the second-order tectonic structure of Olkelduhals, instead of being maximal along SSW–NNE direction, as dictated by the position of the crustal accretion zone (Árnason et al., 2010)?
- why do the earthquakes in the Icelandic crust occur at depths of 12–14 km (Stefansson et al., 1993) where the temperature must have been above solidus?

Analyzing the list of questions not answered by these models, we see that the choice of the conceptual crustal model of the region crucially depends on ability of estimating the spatial distribution of temperature up to a depth of 20–25 km (instead of its estimates at some characteristic depths fulfilled by available indirect geothermometers). Spichak et al. (2013) provided the answers to the questions enumerated above based on (1) estimation of the 3-D temperature distribution in the Hengill geothermal area (Iceland) from the resistivity data up to the depth of 20 km, (2) identification of the heat sources of the geothermal system, and (3) analysis of the seismicity pattern. Below we report the results of this study following above mentioned paper.

8.2   GEOLOGY AND VOLCANIC ACTIVITY IN THE AREA

The earth's crust in Iceland is composed of volcanic rocks with inclusions of intrusive and effusive rocks (mainly oceanic-type flood basalts, tuffs, hyaloclastites, and some felsic rocks). The high-temperature Hengill area is a triple junction zone of intersection of the Western Volcanic Zone (WVZ), the Reykjanes Peninsula Rift (RPR), and the South Icelandic Seismic Zone (SISZ), which is located in the southwest of the island (Einarsson, 2008) (Figure 8.1, upper panel). The Hengill volcanic complex comprises several interconnected geothermal fields located in different directions with respect to the Mt Hengill (marked by H in Figure 8.1, lower panel): the Hveragerdi (Hv) area in the southeast, the Nesjavellir (Ne) area in the northeast, and Hellisheidi (He) area in the southwest (Arnorsson, 1995; Arnorsson et al., 2008; Zakharova and Spichak, 2012).

Overall, the region and its immediate vicinity hosts four centers of volcanic activity: the Hengill area mentioned above, as well as the Grensdalur, Hromundartindur, and Husmuli areas. The Hengill volcanic complex comprises an active central volcano and a swarm of fractures trending north-northeastwards (Figure 8.2). A secondary tectonic structural trend, perpendicular to the dominant

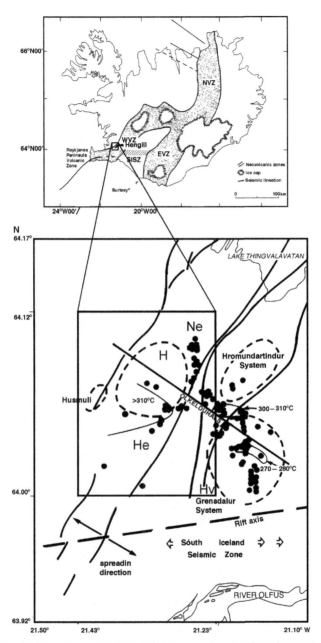

FIGURE 8.1   Upper panel: location of the study area. Lower panel: schematic tectonic map of the Hengill triple junction (modified from (Foulger and Toomey, 1989)). Bold lines indicate the NNE trending eruption/fissure zones. The eruptive centers are outlined by dashed lines. Hot springs and fumaroles are indicated by dots. The line connecting the Hengill and Grensdalur volcanoes indicates the axis of the Olkelduhals transverse tectonic structure. Rectangle bounds the study area, H means Mt Hengill, He – Hellisheidi, Ne – Nesjavellir, Hv – Hveragerdi.

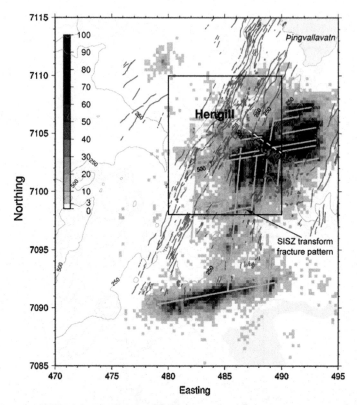

FIGURE 8.2   Density of seismic epicentres (number within 250 m × 250 m bins) from 1991 to 2001 and inferred transform tectonic lineaments (green lines) based on the overall distribution of the seismicity (blue lines: faults and fissures mapped on the surface; red dots: surface geothermal manifestations) (redrawn from Árnason et al. (2010)). Rectangle bounds the study area, and white dashed line indicates location of the Olkelduhals secondary tectonic structural trend shown in Figure 8.1.

NNE–SSW trend of the signs of crustal accretion, has developed in the zone connecting the centers of the Hengill and Grensdalur volcanic complexes and extending along the Olkelduhals line (Figures 8.1 and 8.2). Attached to it are the eruption centers, which migrate west-northwestwards (Foulger, 1988a), as well as the hot springs and fractures.

A vast high-temperature geothermal area that includes the Hengill and Gresdalur central volcanoes as well as the transversal tectonic structure between them is characterized by continuous microseismicity. There is a strong negative correlation between the seismicity and faulting observed at the surface: the earthquakes are clustered around the SN or WNW–ESE azimuths but not in the SSW–NNE direction which dominates the surface geology. On the other hand, the seismicity spatially correlates with heat losses through the surface (Foulger, 1988b), which indicates that seismic activity in this region is associated with geothermal processes rather than with the plate boundary (Foulger, 1988a).

8.3 ELECTROMAGNETIC SOUNDING

8.3.1 Electromagnetic Data

In order to build a 3-D resistivity model of the study area magnetotelluric and a central-loop transient electromagnetic (TEM) data were used (see Figure 8.3 for the sites locations). Figure 8.4 shows a typical example of 1-D inversion of TEM sounding data provided in more than one hundred sites (Árnason et al., 2010). Besides, were used 1-D resistivity profiles revealed from MT data collected in the frequency range from $5 \cdot 10^{-4}$ Hz to 300 Hz at 50 sites with a remote reference point located 10 km apart. The distance between the MT sites and the ocean coast was sufficiently big at this frequency range, so the coast effect on the further MT data interpretation could be neglected (Beblo et al., 1983).

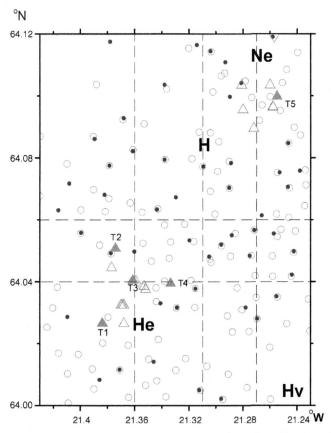

FIGURE 8.3 Map of EM sites and boreholes for the study area indicated in Figure 8.1 (lower panel). Circles mark TEM sites, dots indicate MT sites, triangles mark boreholes (shaded ones mark boreholes used for EM geothermometer testing). H means Mt Hengill, He – Hellisheidi, Ne – Nesjavellir, Hv – Hveragerdi. Dashed profiles indicate projections onto horizontal plane of the vertical temperature cross-sections shown in Figures 8.9–8.10 (redrawn after (Spichak et al., 2013)).

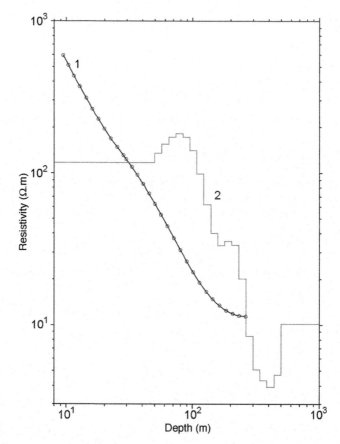

FIGURE 8.4 A typical TEM sounding and its 1-D inversion: 1– measured late time apparent resistivity, 2 – 1-D resistivity model (redrawn from Árnason et al. (2010)).

Figure 8.5 indicates four examples of the apparent resistivity and phase curves (Árnason et al., 2010). According to these authors in the majority of the MT sites the *xy*- and *yx*-components manifest 1-D behavior up to the periods of 1 s (Figure 8.6 indicates the dimensionality indicator "skew" maps for periods ranging from 0.01 to 10 s). So, taking into account that the background resistivity in the Hengill area varies in the range 15–100 Ωm (Árnason et al., 2010; Spichak et al., 2011a) it is reliable to extend the 1-D approximation of the resistivity structure at least to the depth of 5 km.

Unlike the most popular approach to reducing of the so-called "static shift" effect by correcting the MT curves with TEM data measured in the same locations (see, for instance Árnason et al. (2010)), Spichak et al. (2013) used to this end all available TEM and MT data (not necessarily collected in the same sites). In this case, 1-D resistivity profiles determined from the TEM data were used up to the depth of 1 km while 1-D resistivity profiles

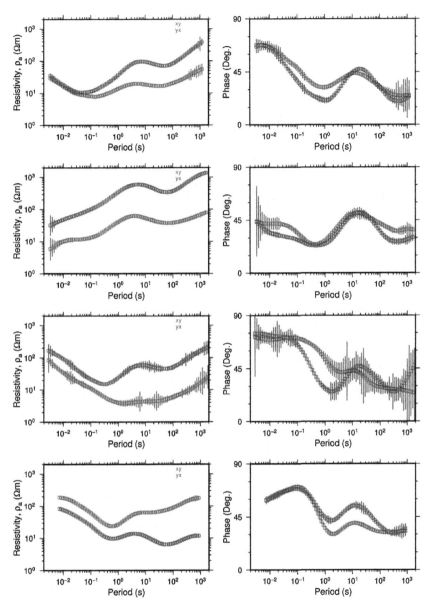

FIGURE 8.5 Four typical examples of apparent resistivity and phase curves for MT sites located in the study area (Figure 8.3) (after Árnason et al., 2010).

determined from the MT data – from 1 to 3 km. This depth limit was justified by two factors: on the one hand, it exceeds the maximal depth of the
temperature logs available in this area but, on the other hand, it is less than
1-D dimensionality depth limit estimated above. So, this procedure enabled

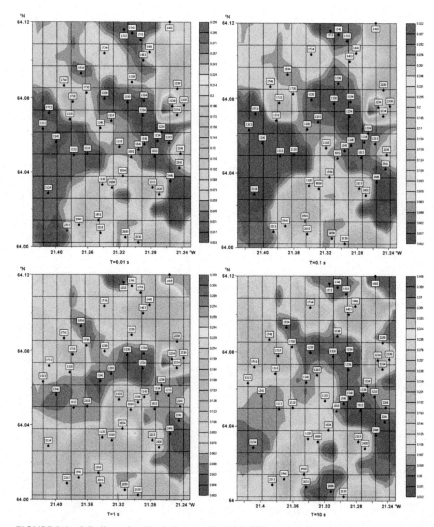

FIGURE 8.6   3-D dimensionality indicator (skew) for different periods.

to avoid the effect of the near-surface geological noise and to enlarge the available database, which, in turn, increased the accuracy of the resistivity reconstruction.

8.3.2   3-D Electrical Resistivity Model

For estimating the temperature in the nodes of the volumetric grid by the electromagnetic geothermometer, we used the 3-D resistivity model of the studied domain which was previously constructed in (Spichak et al., 2011a) at the nodes

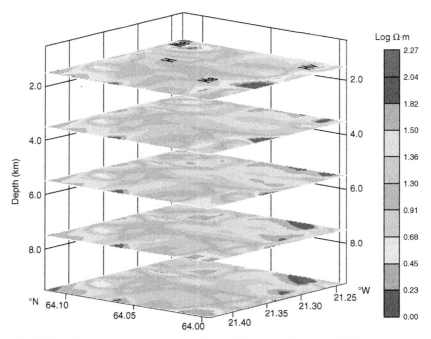

FIGURE 8.7 Horizontal slices of the resistivity model (after Spichak et al., 2011a).

of regular 3-D grid with a step of 0.004° in latitude, 0.005° in longitude, and 2.5 km along the depth (Figure 8.7). Below, we briefly summarize the main conclusions following from the analysis of this model.

The background values of resistivity in the considered domain range from 15 to 40 Ωm. Figure 8.7 shows two anomalies in apparent resistivity, one at the southern boundary of the region and the other, in the northeast. In particular, in the Nesjavellir region between 21.24° and 21.32°W, in the depth interval from 1 to 10 km there is a highly conductive anomalous domain with a horizontal diameter of about 1.5 km and vertical dimension of 2 to 7km, which is rooting to below 20 km. Northeast of Mt. Hengill, in the vicinity of 64.09°N, 21.31°W, this anomalous structure is located at the depths from 4 to 11 km. In this connection, it is remarkable that the results of 2-D inversion of the profile magnetotelluric soundings that were previously conducted in this region were interpreted by Hersir et al. (1984) as a conductive layer with a thickness of 2 km located at a depth of below 10 km.

According to the review by Björnsson et al. (2005) on the MT studies carried out in the region, there are two "highly conductive layers" revealed: the subsurface layer (at about 2 km) and the deep layer (in the depth interval 10–15 km). On the other hand, 3-D resistivity models of this region (Árnason et al., 2010; Spichak et al., 2011a) do not contain continuous conductive layers at these depths. In particular, in the latter paper separate high-conductivity channels interconnected both vertically and horizontally with a resistivity less than 10 Ωm are recognized in the Nesjavellir, Hengill, Hellisheidi, Hveragerdi, and west of Husmuli (Figure 8.7).

In particular, the presence of a highly conductive horizontal "channel" between the Grensdalur area, which is inactive at ptresent, and the currently active Hengill area agrees with the results presented by Árnason et al. (2010), which report on the presence of a highly conductive channels at a depth of 5 km oriented in the direction of the transversal Olkelduhals transform system. These results support the hypothesized W–NW migration of active volcanism from Grensdalur along the Olkelduhals line (Foulger and Toomey, 1989).

A deep conductive fault is identified crossing the study area in the band of 21.31°–21.33°W (Figure 8.7). Its surface projection is located slightly west of the location of the hypothesized transform fault suggested by the surface projections of the earthquake hypocenters (Figure 8.2) (Árnason et al., 2010). At the same time, the horizontal slice of resistivity at the near-surface depth (Figure 8.7) manifests a clear NNE–SSW orientation of the contours of resistivity, which coincides with the predominant trend of the cloud of fractures and outcropping faults (Figure 8.2). This seeming inconsistency is probably accounted for by the right-lateral strike-slip displacement (towards the surface) on the mentioned S–N fault. This displacement results from the interaction between the general field of extension (tensile stresses) acting in the crust of the region, on one hand, and the compressive and shear stresses caused by the ascent of the partially molten mantle material or supercritical fluids, on the other hand (Stefansson et al., 2006).

8.4   EM GEOTHERMOMETER APPLICATION

8.4.1   Temperature Data

Both the high- and relatively low-temperature geotherms recorded in 20 boreholes with different depths and drilled in geologically different Hellisheidi and Nesjavellir geothermal fields were used (see Figure 8.3 for their locations and Figure 8.8 for geotherms indicated by blue lines). The temperature gradient in the boreholes drilled to a depth of 1–2 km varies from 84 ± 9°C/km in the low-temperature regions of the transform zone to 138 ± 15°C/km in the geothermal areas (Foulger, 1995).

Both the geothermal areas of Hellisheidi and Nesjavellir show common characteristics in their measured geothermal gradient as they both are dissected by a volcanic fissure swarm with predominant SSW–NNE trend (Figure 8.2). The southeastern part of the block of boreholes in the Hellisheidi area is mainly controlled by cooling caused by intrusion of cold groundwater to a depth of 1.4–2 km (Franzson et al., 2010). Correspondingly, in most of the temperature logs from this block, the initial temperature rise is followed by a decrease at a depth below 1–1.5 km (Figure 8.8). Within the Nesjavellir geothermal field

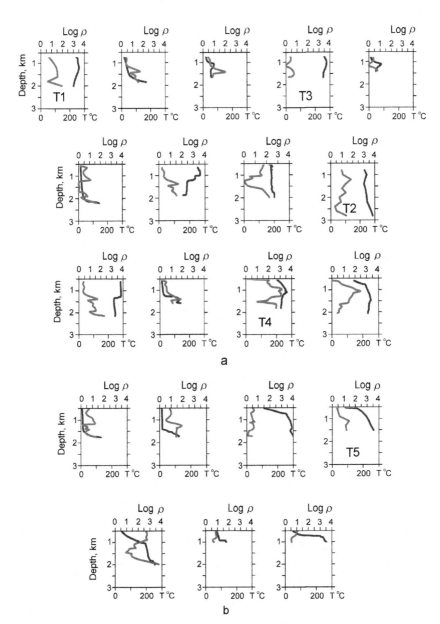

FIGURE 8.8 Temperature well logs (blue lines) and logarithm resistivity profiles (red lines) in the same boreholes: a – Hellisheidi; b – Nesjavellir (after Spichak et al., 2013).

(Figure 8.8b) there is a difference between the temperature patterns of the southwestern, "cooled" part of the region and its eastern "hot" part, probably, caused by intrusion of the overheated vapor through one of the fractures associated with heating in the eastern part of the area.

8.4.2 Calibration

Before training of the artificial neural network (ANN) by correspondence of the resistivity and temperature data, it was necessary first to assign the resistivity values in the locations of the temperature records. (In principle, this step could be avoided if both TEM and MT sites were located close enough to the corresponding boreholes, which rarely happens in practice.) To this end, the first ANN-1 was created and trained for correspondence of the resistivity values yielded by 1-D TEM and MT resistivity data up to the depth 3 km to the coordinates of the appropriate nodes of 1-D profiles where they were determined.

The ANN-1 trained by these resistivity profiles was used then for estimating the resistivity values at locations of the temperature records in 20 wells situated within the Hellisheidi and Nesjavellir geothermal fields. Figure 8.8 indicates the resulting 1-D resistivity profiles forecasted by the ANN-1 at the locations of the boreholes.

Calibration of the EM geothermometer was carried out by another ANN-2, which was actually the indirect EM geothermometer. It was constructed and trained for the correspondence between the values of resistivity and coordinates of the grid nodes, on one hand, and the values of temperature determined in the same nodes, on the other hand. It is important that inputting the coordinates of the grid nodes in addition to the resistivity values to the second neural network implicitly allowed for the dependence of resistivity on the local factors such as the geological and hydrological conditions (Spichak et al., 2011c). This ensured higher accuracy of the temperature estimates compared to those which might have been obtained using the temperature dependences of resistivity determined by laboratory testing of the core samples taken from some boreholes.

8.4.3 Testing

Similarly to the testing procedure applied above in Chapter 7, the temperature and resistivity profiles in each borehole were first linearly discretized (divided into 10 bins). Then for each borehole ANN was trained by the values of two parameters taken from the same locations on 1/10, 2/10, …, 9/10, fractions of the depth of the corresponding profiles. Finally, trained ANNs were used for the temperature forecasting onto the rest parts of the profiles not included in training (9/10, 8/10, …, 1/10). The forecasting errors generally decrease as the ratio (δ) between the borehole length and the extrapolation depth increases (on average, from 26.5 % at $\delta = 0.1$ to 2.7 % at $\delta = 0.9$) (Table 8.1), while starting from $\delta = 0.4$ they are on average less than 5% (Figure 8.9).

If the extrapolation is based on the temperature and resistivity data measured below 500 m, the relative average error of forecasting is practically invariant to the depth of the prediction interval (Spichak and Zakharova, 2009a, 2009b). So, assuming that the errors of extrapolation estimated in a depth range from 0.5 to 2 km remain the same at larger depths, we can roughly estimate the relative temperature prediction errors when the depth

TABLE 8.1 Temperature prognosis errors (%) for five testing boreholes depending on the portion δ of the resistivity and temperature profiles used for calibration

δ	T1	T2	T3	T4	T5
0.1	27.0	22.7	17.0	20.1	45.5
0.2	14.0	28.3	3.4	6.5	16.6
0.3	15.2	10.2	3.6	8.7	4.9
0.4	2.5	4.0	2.1	7.6	2.0
0.5	2.6	2.8	1.6	7.5	1.0
0.6	3.0	4.3	1.5	2.7	3.4
0.7	2.4	5.2	2.8	2.5	0.9
0.8	5.1	3.6	0.5	3.0	0.6
0.9	4.7	4.7	1.2	2.7	0.2

After Spichak et al., 2013.

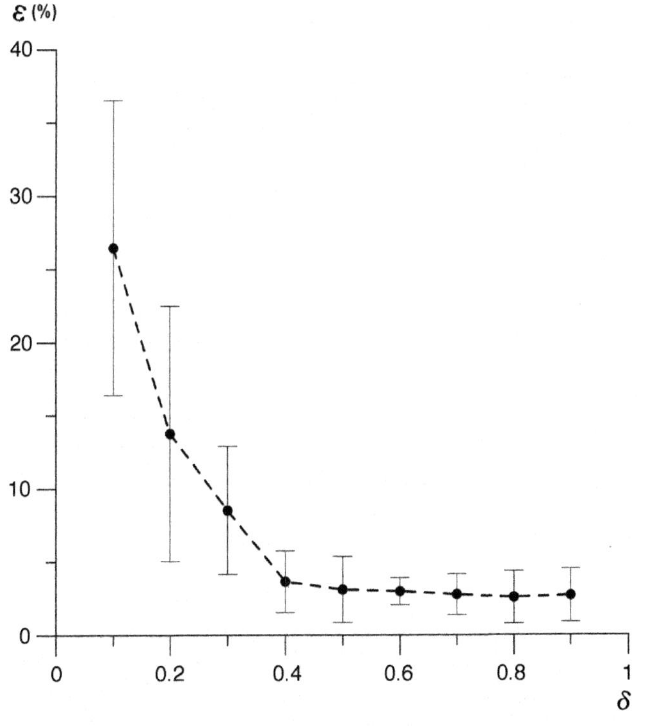

FIGURE 8.9 Dependence of the average relative error ε (%) of the EM temperature extrapolation on the portion δ of the temperature well logs and electrical resistivity profiles used for the neuronet training (after Spichak et al., 2013).

ranges up to 20 km using Figure 8.9. Note that this inference concerns only the temperature assessment itself and is not implicitly based on the preliminary assumptions about the scale similarity of heat conduction mechanisms, geochemical conditions, lithology, and the like.

8.5   3-D TEMPERATURE MODEL

The temperature in the studied area was estimated by Spichak et al. (2011b) by means of the calibrated EM geothermometer (see Section 8.4.2). The analysis was limited to only forecasting temperature up to a depth of 20 km, where the average relative temperature error remains on a level of 26.5%, according to the test calculations described above in Section 8.4.3.

Figure 8.10 indicates the horizontal temperature slices at different depths; the S-N cross-sections are shown in Figure 8.11 while the W–E ones are indicated in Figure 8.12. The isosurfaces of basalt solidus and liquidus (assumed to be equal to 650 and 1100°C, respectively) are displayed in Figure 8.13. Analysis of the spatial temperature distribution presented in Figures 8.10–8.13 may gain an insight into the general pattern and local features of temperature distribution in the region.

8.5.1   Background Temperature

The distribution of the background temperature could be considered as two-layered. The upper layer extends from the surface to a depth of 5–7 km and has a temperature below 200°C. The temperature in the lower layer, located in the depth interval from 5 to 7 km to at least 20 km, varies from 200 to 400°C. Thus, the background temperature above 20 km does not exceed 400°C, which corresponds to the solidus of the silica rich Icelandic gabbro (Wiens, 1993; Menke and Levin, 1994; Fournier, 2007). This conclusion agrees with the gravity data (Darbyshire et al., 2000; Kaban et al., 2002), which support the hypothesis of gabbro and peridotite composition of the lower crust at a temperature below solidus, and is consistent with the temperature estimates made by Hermance and Grillot (1974) based on the solution of the heat transfer equation.

The two-layered background temperature section is overlapped by a network of interconnected high-temperature channels (with the maximum temperature being above 1200°C), which lace through the entire region mainly at depths 10–15 km (Figure 8.13b) and are rooting to a depth below 20 km. Analyzing the horizontal temperature slices (Figure 8.10), we can trace two areas of steep gradients. One zone extends from the south northwards up to the Mt Hengill approximately along the band 21.31–21.33°W, coinciding in its southern part with the location of the deep S–N transform fault hypothesized by Árnason et al. (2010) based on the seismicity pattern (Figure 8.2). The second zone trends parallel to the Olkeduhals transverse tectonic structure (Figure 8.1) and deepens north–northeast from 12.5 to 17.5 km.

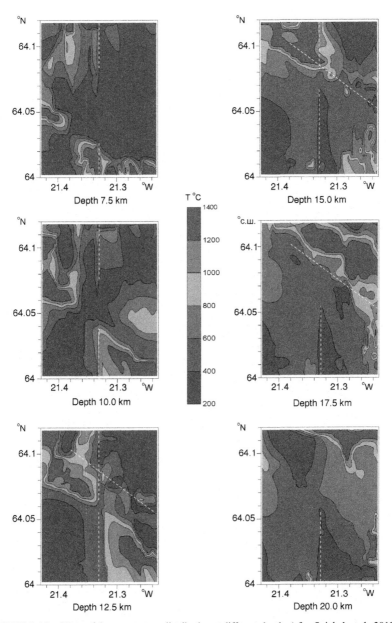

FIGURE 8.10 Slices of the temperature distribution at different depths (after Spichak et al., 2013). Vertical dashed line indicates the hypothesized location of the deep transform fault; diagonal dashed line marks the projection of the Olkelduhals transverse tectonic structure indicated in Figure 8.1.

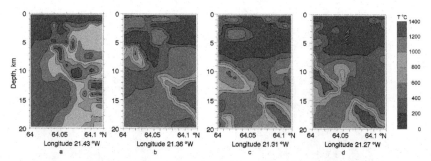

FIGURE 8.11 Temperature cross-sections at different longitudes crossing the western margin of the studied area (a), Hellisheidi (b), Hengill (c), Nesjavellir (d) (after Spichak et al., 2013). White dots indicate the earthquake hypocenters according to (Jousset et al., 2011).

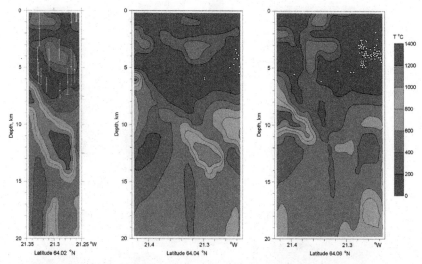

FIGURE 8.12 Temperature cross-sections at different latitudes (see Figure 8.2 for their projections onto the horizontal plane) (after Spichak et al., 2013). White dots indicate the earthquake hypocenters according to (Jousset et al., 2011).

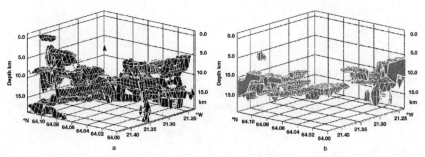

FIGURE 8.13 Isosurfaces of the basalt solidus (650°C) (a) and liquidus (1100°C) (b) (after Spichak et al., 2013).

Another regional feature of the temperature distribution, which follows from the analysis of its S–N cross-sections (Figure 8.11) is the tendency of isotherms to deepen northwards (i.e., off the rift axis that passes south of the considered region (see Figure 8.1 for its location). This temperature behavior is supported by the results of thermomechanical modeling of the crust accretion in Iceland (Schmeling and Marquart, 2008) and the hypothesis of a "thick" crust mentioned above.

The temperature cross-sections shown in Figures 8.11 and 8.12 suggest that the subsurface area at depths 1–5 km contains local anomalies with a diameter of 1.5 to 2 km (the temperature in their central parts ranges from 400 to 600°C) often connected with the high-temperature anomalies beneath them. This pattern of temperature distribution in the shallow subsurface layer agrees with the previously hypothesized penetration of magma to the upper crust and formation of shallow dikes, intrusions, and conduits (Flóvenz and Saemundsson, 1993; Foulger, 1995; Feigl et al., 2000; Árnason et al., 2010) and is consistent with results of seismic tomography of the upper 5 km of the crust (Foulger and Toomey, 1989; Jousset et al., 2010).

Perhaps, the most important characteristics of the temperature distribution in the studied area is the location of the isosurfaces of basalt solidus and liquidus. As seen from Figure 8.13a, the solidus isosurface largely lies at 5–17 km, rising to 5 km in the northwest and southeast of the region (towards the rift), whereas the isosurface of liquidus (Figure 8.13b) splits into two arms. One trends parallel to SE–NW direction at a depth 10–17.5 km beneath the Olkelduhals tectonic structure (see also the temperature sections for these depths in Figure 8.10). Another arm, recognized in the southeast of the region in a depth interval of 7.5–15 km, is localized in the area adjacent to Hveragerdi.

A remarkable fact is that the vertical distance between these isosurfaces (where both of them are recognized) is only 1–2 km. In other words, the temperature gradient in the vicinity of the highly conductive channels may attain 200–400°C/km. (It is worth mentioning that it is close to the temperature gradient estimated at 6–7.5 km depth in the Nesjavellir region from the MT data (Hersir et al., 1984).) On the other hand, it turns to be threefold higher than expected at these depths from the linear extrapolation of the borehole temperature gradient data with the crust tacitly assumed to be lithologically uniform (Björnsson, 2008). For example, the temperature section along 64.02°N (Figure 8.12), which corresponds to the western frame of the SISZ, shows the top of the high-temperature ($T = 1000–1200$°C) layer gradually deepening from 6.5 km in the west to 11 km in the east, which agrees with estimates obtained from seismic data (Hersir et al., 1984; Tryggvason et al., 2002). At the same time, the isotherms $T = 600–800$ °C lie here at a depth of 5.5–10 km, that is, only 1km higher (compare the isosurfaces in Figure 8.13a and b).

8.5.2 Local Temperature Anomalies

It can be seen in Figure 8.10 that an extensive high-temperature anomaly rooting deeper than 20 km in the northeast in the Nesjavellir region extends west, simultaneously ascending to 10–12 km where it joins another high-temperature flow that rises from a depth of 5–7.5 km in the west. As seen from the cross-section along the 21.43°W meridian (Figure 8.11a), at the western margin of the area, below 5 km, there is a branched high-temperature anomaly with $T > 1000°C$, whose slivers extend east and the roots are below 20 km.

From the S–N temperature cross-sections, which intersect the geothermal fields of Hengill, Hellisheidi, and Nesjavellir (Figure 8.11b–d), we see that another high-temperature anomaly frames the region. It rises from a depth of approximately 14 km near Hveragerdi to about 6–8 km close to Hellisheidi. This feature confirms the hypothesis of the heat source in the Hveralio area (located somewhat to the south of Hellisheidi) suggested by Franzson et al. (2010) on the basis of analyzing the geotherms in wells.

Beneath the northern slope of the Mt Hengill at a depth of approximately 10–11 km, an anomaly is identified which has a diameter of about 1.5–2 km and a temperature above the solidus (600–800°C) (Figure 8.11c). This result is consistent with the interpretation of teleseismic data by Foulger and Toomey (1989), which implies a low-velocity body about 5 km^3 in volume to exist below 5 km. This anomaly extends farther north; southwards, directly beneath the Mt Hengill it is identified at a depth of about 12–15 km (see the corresponding slices in Figure 8.10).

Figure 8.12 presents the W–E temperature sections crossing the central part of the area. The local high-temperature anomalies in their eastern margins are likely part of more extensive deep anomaly confined to the Grensdalur geothermal system east of the considered region (its location is shown in Figure 8.1).

8.6 INDICATING HEAT SOURCES

Joint analysis of the temperature and other geophysical data enables to distinguish between the active and relict parts of the geothermal system. To this end, we will consider the spatial distributions of temperature and electrical resistivity, taking into account the gravity anomalies detected in this area by Þorbergsson et al. (1984).

Based on the resistivity model, Spichak et al. (2011a) concluded that the heat sources in the upper crust of the study area could be formed by the upflow of hot highly conductive material (with resistivity less than 10 Ωm) from below 20 km, its accumulation in the subsurface reservoirs and further spreading in the rheologically weak layer at a depth of 5–15 km. The obtained results confirmed the mantle origin of the heat sources in this region, which was hypothesized in (Hermance, 1981; Flóvenz and Saemundsson, 1993; Björnsson et al., 2005).

Spichak et al. (2011a) suggest that the vertical channels could serve as conduits through which liquid magma supposedly wells up from the mantle. Similar plum-like structures were detected recently by MT sounding to SE from the Hengill area (Miensopust et al., 2012) and in the Taupo volcanic zone, New Zealand (Heise et al., 2010; Bertrand et al., 2012). Heise et al. (2010) interpret these vertical channels as zones of interconnected melt rising from depth below 35 km, while in the latter publication they are interpreted as stable convection plums (though this hypothesis is not supported by appropriate temperature estimations at large depths). According to (Shankland and Waff, 1977), the supposed resistivity of 5–10 Ωm and temperature of 1100°C of the highly conductive material correspond to 10–20% melt fraction while according to Gebrande et al. (1980) it ranges between 17 and 23%.

The lacking or incomplete coincidence of highly conductive and high-temperature areas may reflect the dynamics of the thermal process. In particular, the heterogeneous temperature field within the western section of the area (Figure 8.11a) and positive gravity anomaly detected in the corresponding area on the surface (Figure 8.14, I) count in favor of hypothesis (first reported by Foulger and Toomey (1989)) that they could be caused by solidified cooling magma.

Similarly, in the southeastern part of the region close to Hveragerdi a hot zone ($T > 800$°C) is revealed at the depths 10–12.5km (Figure 8.10),

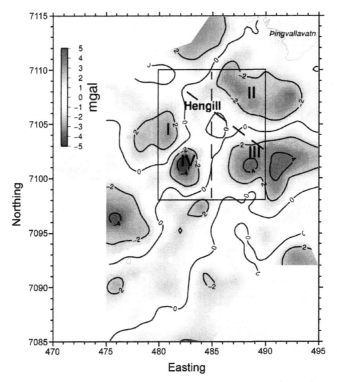

FIGURE 8.14   Residual Bouguer gravity anomaly map (modified after Árnason et al. (2010)); I-IV indicate gravity anomalies; vertical dashed line indicates projection of the deep resistivity and temperature fault; diagonal dashed line marks the axis of the Olkelduhals transverse tectonic structure. Rectangle bounds the study area.

which may indicate the presence of the partially molten magma. On the other hand, moderate values of electrical resistivity (20–30 Ωm) at these depths (Figure 8.7) and positive gravity anomaly in the adjacent Grensdalur area (Figure 8.14, III) enable to suppose that hot magma originated at large depths south from the Hveragerdi area (Franzson et al., 2010) has moved to the north feeding the Grensdalur system, which is presently solidifying and cooling (Foulger and Toomey, 1989).

At the far NE, an extensive high-temperature area is located at large depths (Figure 8.10). The resistivity being rather moderate at these depths becomes less than 10 Ωm at shallow depths (Figure 8.7). Taking into account the negative gravity anomaly (Figure 8.14, II) in this area, this may point to the rock fracturing, which, in turn, provides favorable conditions for forming of the partially molten cooling magma pockets at shallow depths.

Beneath Mt Hengill (approximately at 64.09°N, 21.31°W) a highly conductive vertical anomaly with resistivity 1–3 Ωm and a horizontal diameter of approximately 1 km is located in the depth interval 4–10 km (Figure 8.7). According to the temperature model it has temperatures 400–600°C (Figure 8.10), which may indicate presence of a small volume of partially molten material or supercritical fluids. This is in a good agreement with interpretation of teleseismic data by Foulger and Toomey (1989) who have detected here a low-velocity body about 5 km^3 in volume to exist below 5 km.

The temperature within the low-resistivity zone in the southwestern part of the region (Figure 8.7) is, according to Figures 8.10 and 8.11a, at most 400°C. At the same time, presence of a negative anomaly in the corresponding zone of the gravity anomaly map (Figure 8.14, IV) suggests a fracturing volume at depth, which, in turn, may indicate the possibility of presence of the supercritical fluids of magmatic origin (Stefansson et al., 2006).

Thus, four areas (Figure 8.14, I–IV) discussed above correspond to the regions in the crust with different thermal regimes: in the Husmuli and Grensdalur large massifs of the solidified magma are cooling while in more active Nesjavellir and Hellisheidi areas the upwelling of the partially molten magma could take place. They are separated by a deep S–N fault and the Olkelduhals transverse tectonic structure (marked in Figure 8.14 by dashed lines). The deep fault is traced in the horizontal slices of both electrical resistivity (Figure 8.7) and temperature (Figure 8.10) and coincides with the supposed location of the hypothesized transform fault S–N striking in the southern part of the region (Figure 8.2).

8.7 TEMPERATURE AND SEISMICITY PATTERN

The stability of the spatiotemporal structure of seismicity in the region as well as the fact that seismic activity here is largely confined to the geothermally active areas while almost absent on the plate boundaries (Foulger, 1988a) indicates that seismicity is controlled by geothermal processes which lead to the buildup of local stresses rather than by the tectonic activity caused by spreading.

As seen in Figure 8.2, the maximal concentration of the epicenters (in the horizontal projection) is related to the hypothesized S–N transform faults as well as to the Olkelduhals secondary tectonic structure. According to resistivity model (Spichak et al., 2011a) a deep conductive fault crosses the region in the band 21.31–21.33°W (Figure 8.9). The projection of this fault on the surface practically coincides with the location of the transform fault hypothesized here by Árnason et al. (2000) based on the traced projections of the earthquake hypocenters. At the same time, the dominant trend of the fissure swarm and of the outcropping faults is extended in the NNE–SSW direction. This apparent inconsistency may well be accounted for by the right-lateral thrusting (toward the surface) of the mentioned fault of N–S strike. It could be due to interaction between the tensile stresses acting on the crust in this region, on one hand, and the compressive and shear stresses produced by upwelling of partially molten mantle material, on the other hand (Stefansson et al., 2006). Migration of the partially molten material from large depths to the surface may raise the pore pressure up to lithostatic and thereby enhance instability of the faults.

In the context of our temperature model, such localization of the hypocenters can be related to boundaries between adjacent cooling and heating blocks of the magma (see Section 8.6), which is accompanied by thermal contraction and fracturing, primarily, along the band 21.31°–21.33°W and in a WNW–ESE zone, whose northern boundary follows the Olkelduhals line and the southern boundary is located about 3 km south (see their locations in the temperature slices (Figure 8.10)).

In the vertical plane, most of the earthquakes in the region are known to occur in a depth interval from 2 to 5–6 km. According to the crustal thermal structure discussed above, the high-stress areas could be formed due to cooling in the space between the hot channels, which pass below 5–6 km, and the local reservoirs at a depth of 1–2 km. By correlating the projections of the hypocenters (Stefansson et al., 1993; Jousset et al., 2011) to the vertical temperature cross-sections (Figure 8.11c and d and Figure 8.12), we see that all seismic events are located in the regions where the temperature is below 400°, which is a silica-rich gabbro solidus. This constraint is consistent with the estimates of the critical temperatures of earthquakes (Tse and Rice, 1986; Tichelaar and Ruff, 1993; Wiens, 1993) based on the model calculations for different mechanisms of rock deformation, behavior of the internal friction ratio, and the character of stresses.

The seismicity is located in the areas with electrical resistivity ranging between 15 and 100 Ωm (Spichak et al., 2011a), which could be explained by higher porosity void of melt fractions. It is remarkable that the hypocenters cluster in the regions characterized by both increased and decreased P-wave velocities as well as of Vp/Vs ratio (Figures 8.14 and 8.16, accordingly, from Jousset et al., 2011), which indicates that seismic velocities alone are not indicative of the seismicity pattern. Even joint analysis of the resistivity and seismic velocities data (see, for instance, above mentioned paper) does not always provide enough information, which might enable to draw conclusions on the seismicity origin.

The temperature model also accounts for the occurrence of the earthquakes at large depths (12–14 km and below) reported by Bjarnason and Einarsson (1991). Indeed, the background temperature at this depth does not exceed the silica-rich gabbro solidus. On the other hand, if the true temperature in the location of the earthquake was close to critical (due to closeness to the hot partially melted magma channels upwelling from large depths to the upper crust), then this might have increased the stresses (Stefansson et al., 2006), which, in turn, have raised the probability of the local earthquake to occur.

8.8   CONCEPTUAL MODEL OF THE CRUST

The temperature distribution in the Hengill geothermal area exhibits two layers of the background temperature (one layer in the depth interval from the surface to 5–7 km has a temperature below 200°C; the other layer extending to at least 20 km has temperatures from 200 to 400°C). In this sense, the crust overall can be considered as "thick and cold," which argues for the appropriate conceptual model mentioned in Section 8.1.

On the other hand, the background temperature section which has supposedly gabbro composition is braided by the interconnected highly conductive high-temperature channels – conduits with a diameter of 1–2 km, in the central parts of which the temperature may exceed the basalt solidus. It is important mentioning in this connection that based on the laboratory studies of the electrical conductivity of gabbro Duojun et al. (2002) have concluded that gabbro cannot form any high-conductivity layers in the middle-lower crust. This result implicitly supports our conclusion that the high-temperature low-resistivity channels could consist of partly melted basalt upwelling from the mantle through the crust composed from gabbro at the temperatures below 400°C. It is worth noting in this context that according to (Hermance, 1981) the partially molten basalt could be supplied from the mantle at approximately constant rate through the vertical channels.

According to our model, the heat sources for the geothermal system are probably the intrusions of hot magma composed of partly molten basalt, which rise from the depths of the mantle through the faults and fractures in the Nesjavellir and Hveragerdi fields, south of Hellisheidi (Hveralio) and west of Husmuli. In the former three cases, the magma could directly rise to the rheologically weakened layer at a depth of 10–15 km and then spread laterally. In contrast, in the latter case, the magma supposedly accumulates in the large reservoir, where it then cools down. Such pattern of the crustal structure is consistent with the results of modeling (Schmeling and Marquart, 2008) which showed that, besides the uniform upwelling of magma from the mantle, also the magma chambers and the heat transfer due to geothermal circulation and conductive diffusion are essential elements of the model (Menke and Sparks, 1995).

The interaction between fault tectonics and seafloor spreading, which leads to the formation of fractures and faults, facilitates penetration of hot magma into the permeable upper crustal layers, which, in turn, could result in the emergence of highly conductive shallow dikes and intrusions as well as the magmatic eruptions (Gudmundsson, 1986, 1987; Flóvenz and Saemundsson, 1993). The comparison of these results with those based on MT sounding (Árnason et al., 2010; Gasperikova et al., 2011; Spichak et al., 2011a) suggests that the magma which has risen up to this depth does not form a continuous high-temperature and highly conductive layer, but instead cools down within shallow local pockets at the depths 1.5–2.5 km connected with the underlying hot areas by "chimneys." This agrees with the results of seismic tomography (Foulger et al., 1995), which indicate the local zones with anomalously low shear waves at these depths.

Generally speaking, such a mechanism of forming subsurface intrusions in the Icelandic crust could explain the presence of magma in two wells drilled in Krafla caldera to a depth of 2.6 km with a temperature of 386°C and to a depth of 2.1 km with a temperature 1000°C (Elders and Fridleifsson, 2010). However, according to this paper, instead of basaltic magma they found a rhyolitic one containing ~75% of SiO_2 and less than 2% of H_2O. This disparity is observed also in some Kamchatka volcanoes and could be explained as follows (Anfilogov, 2010). The temperature field around the hot reservoir provides the conditions for Knudsen's molecular diffusion of the pore fluids from the silica rich host rocks (having much less temperature and moderate resistivity ranging in the Hengill case from 15 to 100 Ωm) to much more hot reservoir forming a water saturated melt (supercritical fluid) in the upper part of the reservoir. Its further cooling and partial crystallization could lead to forming of the silica rich rhyolitic melt, which is pulled out from the reservoir to more shallow depths by the basalt magma incoming through vertical channels from the mantle.

8.9 CONCLUSIONS

Thus, application of the new approach to the earth's temperature estimation from the ground EM sounding data enabled to build a new self-consistent conceptual model of the Icelandic crust in the Hengill geothermal area. It agrees with the most of previous geophysical results and provides an explanation for the facts the previous models failed to explain. Answering to the questions stated in the Introduction we could conclude as follows.

Joint analysis of the temperature and resistivity models indicates that highly conductive layers recognized by MT sounding at the depths of 1–3 and 10–15 km are most probably the parts of the hot melted magma network overlapping generally cold Icelandic crust composed from silica-rich gabbroic rocks with solidus equal to 400°C. This could explain why has the drilling in the Krafla geothermal field penetrated rhyolitic magmas with a temperature of $T = 1100$°C at a depth of 2.1 km and with a temperature $T = 386$°C at a depth of 2.6 km.

Joint analysis of the temperature and resistivity models together with the gravity data enabled to reveal the heat sources and discriminate the locations of relict and active parts of the volcanic geothermal complex. This, in turn, explains the observed seismicity pattern by different geothermal regimes in four adjacent parts of the area separated by the deep S-N fault constrained between the meridians 21.31° and 21.33°W and a WNW–ESE diagonal band running beneath the second-order tectonic structure of Olkelduhals.

Above inferences explain the occurrence of the earthquakes in the Icelandic crust at large depths, where, according to the previous crust models, the temperatures have had to exceed the basalt solidus. First, according to the temperature model the background temperatures at least up to the depth of 20 km are less than solidus temperature (400°C) of the crust rocks supposedly composed from the silica-rich gabbro. Second, if the true temperature in the location of the earthquake was close to critical (due to closeness to the hot partially melted magma channels upwelling from large depths to the upper crust), then this might have increased the stresses, which, in turn, have raised the probability of the local earthquake to occur.

REFERENCES

Anfilogov, V.N., 2010. Proiskhozhdenie andezitov i riolitov komplimentarnikh magmaticheskikh serii. Litosphera 1, 37–46 (in Russian with English abstract).

Árnason, K., Karlsdottir, R., Eysteinsson, H., Flóvenz, O., Gudlaugsson, S., 2000. The resistivity structure of high-temperature systems in Iceland (Expanded Abstr). World Geothermal Congress, Kyushu-Tohoku, Japan, pp. 923–928.

Árnason, K., Eysteinsson, H., Hersir, G.P., 2010. Joint 1D inversion of TEM and MT data and 3D inversion of MT data in the Hengill area, SW Iceland. Geothermics 39, 13–34.

Arnorsson, S., 1995. Geothermal systems in Iceland: structure and conceptual models – I. High-temperature areas. Geothermics 24 (5/6), 561–602.

Arnorsson, S., Axelsson, G., Saemundsson, K., 2008. Geothermal systems in Iceland. Icelandic J. Earth Sci. 58, 269–302.

Beblo, M., Björnsson, A., Árnason, K., Stein, B., Wolfgram, P., 1983. Electrical conductivity beneath Iceland – constraints imposed by magnetotelluric results on temperature, partial melt, crust and mantle structure. J. Geophys. 53, 16–23.

Bertrand, E.A., Caldwell, T.G., Hill, G.J., Wallin, E.L., Bennie, S.L., Cozens, N., Onacha, S.A., Ryan, G.A., Walter, C., Zaino, A., Wameyo, P., 2012. Magnetotelluric imaging of upper-crustal convection plumes beneath the Taupo Volcanic Zone, New Zealand. Geophys. Res. Lett. 39, doi:10.1029/2011GL050177.

Bjarnason, I., Einarsson, P., 1991. Source mechanism of the 1987 Vatnafjoll earthquake in South Iceland. J. Geophys. Res. 96, 4313–4324.

Bjarnason, I., Menke, W., Flovenz, O.G., Caress, D., 1993. Tomographic image of the Mid-Atlantic plate boundary in south western Iceland. J. Geophys. Res. 98, 6607–6622.

Björnsson, A., 2008. Temperature of the Icelandic crust: Inferred from electrical conductivity, temperature surface gradient, and maximum depth of earthquakes. Tectonophysics 447, 136–141.

Björnsson, A., Eysteinsson, H., Beblo, M., 2005. Crustal formation and magma genesis beneath Iceland: magnetotelluric constraints. Geol. Soc. Am. Special Paper 388, 665–686.

Darbyshire, F.A., White, R.S., Priestley, K.F., 2000. Structure of the crust and uppermost mantle of Iceland from a combined seismic and gravity study. Earth Planet. Sci. Lett. 181, 409–428.

Duojun, W., Heping, L., Li, Y., Weigang, Z., Congqiang, L., Gengli, S., Dongye, D., 2002. The electrical conductivity of gabbro at high temperature and high pressure. Chinese J. Geochem. 21 (3), 5–12.

Einarsson, P., 2008. Plate boundaries, rifts and transforms in Iceland. Icelandic J. Earth Sci. 58, 35–58.

Elders, W.A., Fridleifsson, G.O., 2010. The Science Program of the Iceland Deep Drilling Project (IDDP): a study of supercritical geothermal resources (Expanded Abstr). World Geothermal Congress, Bali, Indonesia.

Feigl, K.L., Gasperi, G., Sigmundsson, F., Rigo, A., 2000. Crustal deformation near Hengill volcano, Iceland 1993-1998: coupling between magmatic activity and faulting inferred from elasic modeling of satellite radar interferograms. J. Geophys. Res. 105 (B11), 25,655–25,670.

Flóvenz, O.G., 1985. Application of subsurface temperature measurements in geothermal prospecting in Iceland. J. Geodyn. 4, 331–340.

Flóvenz, O.G., Saemundsson, K., 1993. Heat flow and geothermal processes in Iceland. Tectonophysics 225, 123–138.

Foulger, G.R., 1988a. Hengill triple junction, SW Iceland. 1, Tectonic structure and the spatial and temporal distribution of local earthquakes. J. Geophys. Res. 93, 493–506.

Foulger, G.R., 1988b. Hengill triple junction, SW Iceland. 2, Anomalous earthquake focal mechanisms and implications for process within the geothermal reservoir and at accretionary plate boundaries. J. Geophys. Res. 93, 507–523.

Foulger, G.R., 1995. The Hengill geothermal area, Iceland: variation of temperature gradients deduced from the maximum depth of seismogenesis. J. Volcanol. Geotherm. Res. 65, 119–133.

Foulger, G.R., Du, Z., Julian, B.R., 2003. Icelandic-type crust. Geophys. J. Int. 155, 567–590.

Foulger, G.R., Miller, A.D., Julian, B.R., Evance, J.R., 1995. Three-dimensional V_p and V_p/V_s structure of the Hengill triple junction and geothermal area, Iceland, and the repeatability of tomographic inversion. Geophys. Res. Lett. 22, 1309–1312.

Foulger, G.R., Toomey, D.R., 1989. Structure and evolution of the Hengill-Grensdalur volcanic complex, Iceland: geology, geophysics, and seismic tomography. J. Geophys. Res. 94 (B12), 17511–17522.

Fournier, R., 2007. The physical and chemical nature of supercritical fluids. Proc. Workshop on Exploring high temperature reservoirs: new challenges for geothermal energy, Volterra, Italy.

Franzson, H., Gunnlaugsson, E., Árnason, K., Saemundsson, K., Steingrimsson, B., Hardarson, B.S., 2010. The Hengill geothermal system, conceptual model and thermal evolution (Expanded Abstr). World Geothermal Congress, Bali, Indonesia.

Gasperikova, E., Newman, G., Feucht, D., Árnason, K., 2011. 3D MT characterization of two geothermal fields in Iceland. GRC Trans. 35, 1667–1671.

Gebrande, H., Miller, H., Einarsson, P., 1980. Seismic structure of Iceland along the RRISP-77 profile. J. Geophys. 47, 239–249.

Gudmundsson, A., 1986. Formation of crustal magma chambers in Iceland. Geology 14, 164–166.

Gudmundsson, A., 1987. Formation and mechanisms of magma reservoirs in Iceland. Geophys. J. R. Astr. Soc. 91, 27–41.

Heise, W., Caldwell, T.G., Bibby, H.M., Bennie, S.L., 2010. Three-dimensional electrical resistivity image of magma beneath an active continental rift, Taupo Volcanic Zone, New Zealand. Geophys. Res. Lett. 37, doi: 10.1029/2010GL043110.

Hermance, J.F., 1981. Crustal genesis in Iceland: Geophysical constraints on crustal thickening with age. Geophys. Res. Lett. 8, 203–206.

Hermance, J.F., Grillot, L.R., 1974. Constraints on temperatures beneath Iceland from magnetotelluric data. Phys. Earth Planet. Int. 8, 1–12.

Hersir, G.P., Björnsson, A., Pedersen, L.B., 1984. Magnetotelluric survey across the active spreading zone in southwest Iceland. J. Volc. Geoth. Res. 20, 253–265.

Jousset, P., Haberland, C., Bauer, K., Árnason, K., 2010. Detailed structure of the Hengill geothermal volcanic complex, Iceland, inferred from 3-D tomography of high-dynamic broadband seismological data (Expanded Abstr). World Geothermal Congress, Bali, Indonesia.

Jousset, P., Haberland, C., Bauer, K., Árnason, K., 2011. Hengill geothermal volcanic complex (Iceland) characterized by integrated geophysical observations. Geothermics 40, 1–24.

Kaban, M.K., Flóvenz, O.G., Palmason, G., 2002. Nature of the crust-mantle transition zone and the thermal state of the upper mantle beneath Iceland from gravity modeling. Geophys. J. Int. 149, 281–299.

Menke, W., Brandsdottir, B., Einarsson, P., Bjarnason, I.T., 1996. Reinterpretation of the RRISP-77 Iceland shear-wave profiles. Geophys. J. Int. 126, 166–172.

Menke, W., Levin, V., 1994. Cold crust in a hot spot. Geophys. Res. Lett. 21 (18), 1967–1970.

Menke, W., Levin, V., Sethi, R., 1995. Seismic attenuation in the crust at the mid-Atlantic plate boundary in south-west Iceland. Geophys. J. Int. 122, 175–182.

Menke, W., Sparks, D., 1995. Crustal accretion model for Iceland predicts cold crust. Geophys. Res. Lett. 22, 1673–1676.

Miensopust, M., Jones, A., Hersir, G., Vilhjálmsson, A., 2012. The resistivity structures around and beneath the Eyjafjallajökull volcano, southern Iceland: first insights from electromagnetic investigations (Expanded Abstr). XXXIV International Geological Congress, Brisbane, Australia.

Pavlenkova, N.I., Zverev, S.M., 1981. Seismic modeling of the Icelandic crust. Geologischau Rundschau 70, 271–281.

Schmeling, H., Marquart, G., 2008. Crustal accretion and dynamic feedback on mantle melting of a ridge centered plume: the Iceland case. Tectonophysics 447, 31–52.

Shankland, T.J., Waff, H.S., 1977. Partial melting and electrical conductivity anomalies in the upper mantle. J. Geophys. Res. 82, 5409–5417.

Spichak, V., Goidina, A., Zakharova, O., 2011a. Trekhmernaya geoelktricheskaya model vulkanicheskogo kompleksa Hengil (Islandiya). Herald of KRAUNZ 1 (19), 168–180 (in Russian with English abstract).

Spichak, V., Zakharova, O., 2009a. Electromagnetic temperature extrapolation in depth in the Hengill geothermal area, Iceland (Expanded Abstr). XXXIV Workshop on Geothermal Reservoir Engineering, Stanford University, Stanford, USA.

Spichak, V., Zakharova, O., 2009b. The application of an indirect electromagnetic geothermometer to temperature extrapolation in depth. Geophys. Prosp. 57, 653–664.

Spichak, V., Zakharova, O., Goidina, A., 2011b. 3D temperature model of the Hengill geothermal area (Iceland) revealed from electromagnetic data (Expanded Abstr). XXXVI Workshop on Geothermal Reservoir Engineering, Stanford University, USA.

Spichak, V.V., Zakharova, O.K., Goidina, A.G., 2013. A new conceptual model of the Icelandic crust in the Hengill geothermal area based on the indirect electromagnetic geothermometry. J. Volcanol. Geotherm. Res. 257, 99–112.

Spichak, V., Zakharova, O., Rybin, A., 2011c. Methodology of the indirect temperature estimation basing on magnetotelluruc data: northern Tien Shan case study. J. Appl. Geophys. 73, 164–173.

Stefansson, R., Bodvarsson, R., Slunga, R., Einarsson, P., Jacobsdottir, S., Bungam, H., Gregerson, S., Hjelme, J., Kerhonen, H., 1993. Earthquake prediction research in the South Iceland Seismic Zone and the SIL Project. Bull. Seismol. Soc. Am. 83, 696–716.

Stefansson, R., Gudmundsson, J.B., Roberts, M.J., 2006. Long-term and short-term earthquake warnings based on seismic information in the SISZ. In: VeÐurstofa Íslands – GreinargerÐ, Icelandic Meteorological Office, Rep. 06006, 53pp. Reykjavik.

Tichelaar, B.W., Ruff, L.J., 1993. Depth of seismic coupling along subduction zones. J. Geophys. Res. 98 (B2), 2017–2037.

Tryggvason, A., Rognvaldsson, S.Th., Flóvenz, O.G., 2002. Three-dimensional imaging of P- and S-wave velocity structure and earthquake locations beneath Southwest Iceland. Geophys. J. Int. 151, 848–866.

Tse, S.T., Rice, J.R., 1986. Crustal earthquake instability in relation to the depth variation of frictional slip properties. J. Geophys. Res. 91 (B9), 9452–9472.

Wiens, D.A., 1993. Too hot for earthquakes? Nature 363, 299–300.

Zakharova, O.K., Spichak, V.V., 2012. Geothermal fields of Hengill volcano. Iceland. J. Volc. Seism. 6 (1), 1–14.

Ðorbergsson, G., Magnusson, I., Gunnarsson, Á., Johnsen, G., Björnsson, A., 1984. Geodetic and gravity surveys in the Hengill area 1982 and 1983. Orkustofnun Rep. OS-84003/VOD-03 B, Reykjavik, Iceland, 58 pp. (in Icelandic).

Concluding Remarks

Electrical resistivity of rocks could serve as a convenient proxy parameter to be used for indirect estimation of the temperature from the electromagnetic sounding data. In particular, the deep temperature estimation could be based on the indirect electromagnetic geothermometer, which could be used for the temperature assessment in the locations where the resistivity data are available (in particular, revealed from the magnetotelluric sounding data). Unlike other indirect geothermometers it enables the temperature estimation in the given locations in the earth (in particular, at large depths) and building of 2-D or 3-D temperature models of the studied areas.

The temperature interpolation in the interwell space using EM sounding data results in a sharp increase of the estimation accuracy the latter being controlled by four factors: faulting in the space between the place where the temperature profile is estimated and related EM site, distance between them, meteoric and groundwater flows, and lateral geological inhomogeneity.

The accuracy of the EM temperature extrapolation in depth depends only on the ratio between the borehole length and the extrapolation depth (assuming that the EM sounding site is located near to the appropriate borehole). This, in turn, opens up the opportunity to use available temperature logs for estimating the temperatures at depths 3–10 km without extra drilling. This makes the EM geothermometer particularly attractive in prospecting for deep-seated sources of geothermal energy as well as hydrocarbon deposits characterized by temperature anomalies.

Alternatively, the temperature at large depth could be "Forecasted While Drilling" based on the EM sounding data, current temperature logs, and core samples. This gives an impetus to develop in future a new strategy of the deep exploration drilling, which could lead to increasing of the accuracy of the target parameters' estimation and essential reducing of the drilling expenses (EM sounding costs could be neglected since they are two orders less).

In this context, it is important that EM geothermometer is also helpful for narrowing the area of uncertainty when selecting the locations for drilling of new exploration boreholes. The positive experience of such EM geothermometer's application in the Soultz-sous-Forêts geothermal area (France) could be expanded to the other geothermal regions.

The methodological studies give hope that the practical application of the EM geothermometry will make it possible to estimate the temperature in the

Electromagnetic Geothermometry. http://dx.doi.org/10.1016/B978-0-12-802210-8.00009-5

geothermal reservoirs during their exploitation from the regular EM sounding data (i.e., to carry out 4-D temperature monitoring).

The application of the EM geothermometer for estimating the temperatures in the different geological conditions has shown that it is possible not only to construct the models of deep temperature distributions in the studied regions but also, based on the analysis of these models, to make the conclusions concerning the predominant mechanisms of heat transfer at large depths and the locations of the probable heat sources. Our investigations in the Hengill geothermal zone (Iceland) have demonstrated that for constructing the conceptual models of geothermal areas it is very useful to carry out a joint analysis of the temperature and electrical resistivity models taking into account the gravity anomalies.

In a whole, the EM geothermometry offers tools, which help to overcome the bottlenecks of the geothermal industry and thus to increase its competitiveness in comparison with other energy sectors. Note that the accuracy of temperature estimates provided by EM geothermometer depends on the mode of its use (interpolation, extrapolation, monitoring, or construction of the deep models), each of which has its own governing factors. Although our first studies suggest that the obtained temperature estimates are sufficiently robust to the accuracy of the input resistivity data, this question should be treated with a caution: both the selection of the adequate EM sounding technique and the subsequent construction of the resistivity model should be carried out very thoroughly. And the last but not least: the effectiveness of the EM geothermometry depends on using of adequate ANN software tools, self-adapting ones being, probably, the best.

Index

Printed in the United States
By Bookmasters